Exile Politics during the Second World War

The German Social Democrats in Britain

BY

ANTHONY GLEES

CLARENDON PRESS · OXFORD
1982

Oxford University Press, Walton Street, Oxford OX2 6DP

London Glasgow New York Toronto
Delhi Bombay Calcutta Madras Karachi
Kuala Lumpur Singapore Hong Kong Tokyo
Nairobi Dar es Salaam Cape Town
Melbourne Auckland
and associate companies in
Beirut Berlin Ibadan Mexico City

Published in the United States
by Oxford University Press, New York

British Library Cataloguing in Publication Data

Glees, Anthony
    Exile politics during the Second World War: the German Social Democrats in
    Britain.—(Oxford historical monographs)
    1. Sozialdemokratische Partei Deutschlands—History   2. World War,
    1939–1945—Great Britain.
    I. Title
    940.53'22'41   D809.G3
    ISBN 0–19–821893–1

Library of Congress Cataloging in Publication Data

Glees, Anthony, 1948-
    Exile politics during the Second World War. (Oxford historical monographs)
    Bibliography: p.
    Includes index.
    1. World War, 1939–1945—Governments in exile. 2. World War, 1939–1945—
    Germany. 3. Germany—Politics and government—1933–1945. 4. Great Britain—
    Foreign relations—Germany. 5. Germany—Foreign relations—Great Britain.
    6. Sozialdemokratische Partei Deutschlands—History. 7. Labour Party (Great
    Britain)—History. 8. Germans—Great Britain—History—20th century. 9. World
    War, 1939–1945—Diplomatic history.
    I. Title.
    D810.G6G56   940.53'22'43   82–6377
    ISBN 0–19–821893–1   AACR2

Typeset by Graphic Services, Oxford
Printed in Great Britain
at the University Press, Oxford
by Eric Buckley
Printer to the University

Preface

ALTHOUGH responsibility for this book is mine alone, it could not have been written without the help of a number of people and institutions. First and foremost, thanks are due to A. J. Nicholls. It is impossible to conceive of a better supervisor. Fritz Heine, now of the Friedrich-Ebert-Stiftung and a senior member of the SPD, for many years on its Executive and its press chief, who worked closely with Kurt Schumacher and Erich Ollenhauer, offered me many insights into the SPD's development and into our common interest, Social Democracy. Professor James Joll, Dr Peter Pulzer, and Mr A. F. Thompson were kind enough to help me in a number of ways and I am also grateful to the people who allowed me to interview them. Their names are recorded at the back of this book.

Institutions whose assistance I have benefited from are: the Public Record Office, the archives of the Labour party, the library of the London School of Economics, the German Historical Institute, and the Wiener Library, all in London; the archives of the Friedrich-Ebert-Stiftung in Bonn, W. Germany; and the libraries of St Antony's College and University College, Oxford. I received further help from the University of Oxford (in the form of a Michael Foster Memorial Scholarship) and from the German Academic Exchange Service. I must also thank the Department of History at Warwick University and the Department of Government at Brunel University. Without posts at these places, there would have been no book at all.

Finally, I owe my greatest debt to my wife Ann, for her cheerful encouragement and her clear mind. This book is for her.

Anthony Glees,
Department of Government,
Brunel University
September 1981

Contents

Introduction

THE impact made by Hitler and National Socialism on the political development of the nations of modern Europe has rightly received close scrutiny. Almost everyone in the contemporary world was in some way affected by the ambitions of this man and his followers. But it was not only individuals whose lives were altered by Nazi ideas and policies. Political systems and political parties were also changed as a result of the seemingly overwhelming power of National Socialism. The apparent success and the ruthlessness of Hitler's hold on the German nation meant that all those in any opposition to him and to his political settlement, were forced onto the defensive. Parliamentary democracy was under attack and the parties which supported it both inside Germany before 1933 and outside it thereafter, were made to face a terrifying challenge. They could choose whether to abdicate or whether to fight, to surrender to the Nazis or to confront them.

The political dimensions of this struggle against Hitler extend far and wide. A variety of issues are encompassed within them, ranging from the appeasement by some in the thirties to the division of the Germany and the Cold War in the late forties. Scholars have tried to investigate them in a number of ways. There are those who have looked at the problem from the viewpoint of the Third Reich itself, examining matters like the use of terror and even Hitler's alleged psychopathy. Others have tried to assess the decisions made by the great Allied leaders, Churchill, Roosevelt, and Stalin. But a number of questions remain unanswered and they are questions central to the politics of modern Europe.

Many of these concern either directly or indirectly the political history of the Democratic Socialist movement in Germany, the SPD. Hitler's first aim in politics was to destroy the Weimar Republic. Although it was born out of Germany's defeat in the First World War, it was nonetheless the child of the SPD and, despite many of its flaws, Weimar Germany was sufficiently liberal and progressive to incur the wrath of a would-be dictator. After seizing power Hitler attempted to maintain his regime by a number of totalitarian and occasionally pseudo-democratic devices which were designed as radical alternatives to the political ideals of Social Democracy. Hitler strove to disrupt the peace of Europe, to create a New Order built on crazy notions of racial superiority utilizing terror and brute force instead of civilized, democratic persuasion. All of this was conceived as, and carried out as, the direct antithesis of the aims and plans of the SPD.

But German Social Democracy is part of the structure of modern European politics for another set of reasons. First and foremost is the fact that it survived the war. Hitler was destroyed as was his Nazi organization. The SPD, on the other hand, continued to live and see better days. Indeed, the continuity of the Social Democratic party has a double significance. It illustrates the resilience of its ideals and policies and it also sheds light on the nature of post-war German politics. For a central feature of the Federal Republic of Germany has been the loyalty towards it shown by the SPD even though it was in opposition for almost two decades. If Bonn is not Weimar, it is in large part because Bonn possessed a democratic opposition which was committed to the idea of a German democracy. Weimar, however, was plagued by the fact that those opposed to the state were also opposed to the concept of parliamentary democracy.

There is no doubt that the SPD's loyalty to democracy in 1945 has been taken for granted. Few have questioned it, simply because the existence of Social Democracy after 1945 is seen as something quite unremarkable. That this is so is in itself remarkable, for the reasons why there was an SPD in 1945 are by no means straightforward, not least because so much of what Hitler did was directed against the SPD. In addition, the terrible crimes of Nazism not only appeared to warrant a wholly new beginning for German politics after Hitler, but they also appeared to shed grave doubts on whether the German people could ever act democratically again. Did the Germans deserve to run their own political affairs? Would democratic rights not be used once again to destroy democracy? Was it not better to spend years forcibly re-educating Germans who might have been ineradicably tainted by Nazism? Should not German politics be under non-German control, for one generation at least?

But the SPD, the phoenix of German politics, rose again. The continuity of party identity was preserved. And in this process, exile played a vital role. Exile, which lasted from 1933 until the end of 1945, was the means by which the party stayed alive. Exile, as a political process distinct from any other, thus needs close and detailed analysis. This may seem surprising. In many ways the SPD's exile appears to be a problem on the fringes of historical relevance. The number of Social Democrats in exile was very small compared to the mass membership of the party at home. It was out of touch with those who might have otherwise supported it and it was obviously quite incapable of exerting any real influence on German politics until after Hitler's defeat. A party which had until 1933 relied on its mass membership for its power base and its political legitimation might appear not to merit any serious consideration when it was down and out in Prague, Paris, and London.

It is therefore important to state at once that the SPD *was* treated with seriousness in exile in Britain, at any rate until 1943. Although it was ignored thereafter, this was not because it was considered unimportant but because to take account of it was to raise a number of issues which British authorities preferred not to pursue. Certainly, either explicitly or implicitly, the German Labour movement was believed to possess a political future simply by virtue of its position as spokesman of at least a part of the German working class. The view that exile politics are of marginal significance—which a hasty glance at the period from 1941–5 might seem to confirm—is seen to be totally incorrect when the evidence is subject to closer examination.

Such close examination of the SPD in London reveals a number of interesting factors. They range from narrow matters of detail to very broad problems such as the nature of a political party and the way in which policies are generated. The first detail to note is the courage displayed by the exile leaders of the SPD. This took the form both of personal and of political bravery. Here were men and women who refused to go under and were pledged to fight for the Allied cause against Hitler, even to the extent of working for the Secret Services of the British and American governments. The second factor to emerge is the dramatic change in the attitude of the British Labour party and the Foreign Office towards German political exiles that occurred during the course of the war. Although this change was opposed by a minority of people in both institutions, it nonetheless set the seal on subsequent political relationships. Third, we can see evidence of the tenacity of the leadership and its single-minded desire to achieve political power after Hitler.

As far as the broader issues are concerned, there is much evidence to show that the political development of the SPD in exile is not marginal to the political development of modern Europe after 1945, but is very much a central part of it. To begin with, the chiliasm which forms the main plank of Socialist ideology was a vital factor in keeping the SPD intact and living. In crude terms, the SPD believed that the political present, namely the defeat of National Socialism, and the political future, namely the realization of Social Democratic goals, were both part of the same *continuum* of political behaviour. A more pragmatic view of politics would not have enabled the SPD to survive as a separate political entity. A political party, we may deduce, needs its ideology and cannot be divorced from it. In addition, the formation of leadership élites, which exile stimulates, is as important for party development as the formation of a mass membership.

The SPD's enforced exile is also intimately related to the nature of the alleged opposition to Nazism and Hitler. Was there such a thing as the 'Other Germany', a Germany of anti-Nazis, who were unable to express their hostility towards those who governed them? What forms could opposition to Hitler's totalitarian dictatorship realistically be expected to take given that any open acts of opposition would probably lead to torture, incarceration, and even death? So is it fair to argue that because open opposition to the Nazis' regime was such a rare event, there was no real opposition to it? Might it not be the case that the oppositional tendencies of the German people were very much greater than it has commonly been supposed? This was, of course, the view of the SPD in exile which consistently claimed that in a totalitarian state, the true political beliefs of the population were impossible to assess and that silence or even obedience did not imply complicity. On the other hand, the reluctance of German citizens to defy the Nazi authorities, to interfere with Nazi policies of extermination and warfare, and their failure to offer popular support to people like Elser, who attempted to assassinate Hitler in 1938, or the 'White Rose' group after 1943, must inevitably raise serious doubts about the real political strength of the 'other' Germany.

Political exile does not take place in a vacuum. This is true even if the host nation decides after a time, to ignore the exiles —the fate, broadly speaking, that befell the SPD in London. Straightforward permission to reside and the provision of the wherewithal to live are in themselves politically vital supports for exiled politicians. Yet its relationship with the British Labour party and the Foreign Office certainly was the pivotal issue for the SPD in England. The history of this relationship is at least half the history of exile, and to offer an analysis of it also involves an examination of important parts of British policy during the Second World War.

It is perfectly true that the SPD's problems were not ones which kept the British Prime Minister or the Leader of the Labour Party awake at night. There is no record of the SPD ever having been a subject for discussion in the War Cabinet. The formulation of policy towards the SPD was a matter for the second division of political officials. This does not, however, mean the SPD's existence in exile was unimportant or that the questions that it raised had little or nothing to do with the pursuit of a successful policy in wartime and for the period thereafter. Nonetheless, to have considered how German Social Democrats might have participated in the war effort or in the formulation of post-war policies towards Germany, was just not on the agenda of British politics after 1943. This is, in itself, a comment on the sort of agenda that British authorities felt constrained to produce. For different reasons—reasons which examination of the SPD in exile does much to clarify—neither the British government nor the British Labour party wished, after 1943, to devote a great amount of thought to the kind of plans that would have involved close collaboration with the SPD in exile.

Before 1943 the British Labour party believed that the Second World War was, in essence, caused by reactionary political forces and served the interests of capitalism. It thus wished to fight for a Socialist Europe, a post-war system in which Socialist governments would co-operate in order to make a lasting peace and abolish the social, political, and economic reasons for war. In such a plan, German Social Democrats were bound to play a key role, not least because it was largely, though not exclusively, the German crisis which had led to war. After 1943, however, the British Labour party took a different line altogether. Those who wished to see a Socialist Europe were considered 'softliners' towards the Nazis—pacifists and left wing idealists. There had been a willingness to work with the SPD as long as it agreed to act as the agent of British Labour. But because it had asserted its independence, it was to be ignored and there were hopes that it would disappear altogether. The fact that, working together, the British Labour party and the SPD might have been able to construct a better post-war world founded on Socialist solidarity was quietly forgotten.

In the same way that pro-SPD policies became unsavoury in the British Labour movement, policies which involved co-operating with German exiles also became redundant in the British Foreign Office. Here, too, a change took place, a change which has not received sufficient attention in the past. In the first two or three years of the war, the Foreign Office tried very seriously to weave German political exiles into the thread of its policies towards Germany. It saw in them a useful source of information and also a possible nucleus of an alternative anti-Nazi movement which might make peace with Britain and lead to the defeat of Hitler without the necessity of lengthy conflict and consequent loss of life. Yet with the arrival of Anthony Eden in the Foreign Office, a slow but sure change in British policy took place, one which led to the exclusion of this exile component. Any desire to stop short of the outright destruction of Germany and of unconditional surrender smacked of appeasement policies. In addition, after 1941, there was a Russian dimension. It was thought that any move towards the German opposition might be seen in Moscow as a move towards a separate anti-Bolshevist peace. It was, of course, important that the Red Army carry on fighting. But the fact remains that the Soviet government pursued a very different policy towards German exiles which the British authorities failed to comprehend. And it was one which led ultimately to the creation of a Communist stranglehold on part of Germany which remains intact to this day.

These were the challenges that the SPD had to face in exile in addition to the challenge of attempting to survive the Nazi years of power. But there was one final contest. And that was with the Communists and the KPD. By 1943 relations first with the Foreign Office and then with the British Labour party had ground to a halt. The SPD was at its weakest and very close to complete disintegration. At this point approaches were made by the KPD leaders in London with the intention of amalgamating the two Socialist parties in Germany after Hitler had been defeated, thus producing a united political wing of the German Labour movement. The enormous prestige of the Soviet Union was thrown behind this proposal and many Social Democrats in exile were tempted to join such a unified Free German Movement. But the SPD stood firm against it. It was ironic that this firmness, which precipitated the gravest exile crisis for the SPD, in fact did more than anything else to ensure the SPD did not go under. For by 1944 it became clear that the Soviet Union wished to control parts of German territory and that the Free German Movement was expected to support the destruction of German national unity. The SPD was thereby able to show that it had not only always opposed the Communists but had fought for Germany's independence. It could retain both a Social Democratic character and speak for the German nation from the left of the political spectrum.

In this way, the SPD was able to stay in the business of politics although many objective factors militated against it. It was able to preserve its identity and its integrity. It was able to avoid repeating the mistakes of 1918, when it had not been able to assert its difference from the regime which had preceded it. It could adopt a special view of Germany and of the sorts of policies it believed a new Germany would require. They were nationalistic but unaggressive; anti-Communist but not irrational. Above all, the SPD was able to provide the fledgling Republic with an opposition that was completely loyal to the idea of Republican Democracy and thus enable it to overcome the initial crises that were to face it. The sort of state that West Germany has become is in large part the result of the SPD's view of itself and the SPD's view of itself is in large part the outcome of the harsh years that it spent in exile.

PART I
1933–1941

I

The Background to the SPD's Exile in Britain: the Period from 1933 to 1941

IN order to demonstrate that German Social Democracy did not die during or before the summer of 1933, it has to be proved that the SPD leadership, or the Sopade leadership as it termed itself after the flight from Germany, was able to execute two fundamental tasks. The first was the preservation of a distinctive social democratic identity which had to be achieved through legitimation. The second was a real and an unromantic assessment of the likely requirements of the SPD's future political constituency with whom they were no longer in direct contact. In both instances the historical traditions of the SPD which had included a period of exile from 1878 until 1890 were bound to play a central role.

Somewhat surprisingly, perhaps, an examination of the existing evidence concerning the Sopade's sojourn in Prague and Paris indicates that these two functions were not carried out with any consistency. It was only when the decision to flee to England was taken that the SPD leadership gained a new seriousness. Until 1940, then, German Social Democracy in exile was on the verge of complete disintegration.

It should not be supposed that the Sopade's problem was the result of exile *per se*. Exile is by no means synonymous with political extinction. And, indeed, the opposite can be the case.[1] General de Gaulle, for example, had no popular following in France in 1940 yet he emerged from exile as a leading French political figure whose subsequent career was remarkable. In an earlier era Lenin had been able to emerge from exile to assume the leadership of the Russian revolution and become the first Bolshevik head of state.

Exile, therefore, is not to be seen as inevitably the end to

[1] Recent events in Iran confirm this view.

any domestic political career. But it is clear that if it is to be successful then the exiles themselves must in some way meet the challenges of their position. Neither in Prague nor in Paris did German Social Democratic leaders manage to demonstrate that they were able to exercise effective leadership over their followers in Germany.

This, it will be argued, was unlike the SPD's activities in London, where to a greater or lesser extent a different and younger leadership did manage to achieve legitimation from within, to gain a measure of recognition from the British authorities, and to demonstrate realism in assessing the SPD's return route to German politics. Even if these external relationships rapidly deteriorated after 1941, its internal relationships and its ideological standpoint became firmer. Men like Erich Ollenhauer, Willi Eichler, and Fritz Heine injected dynamism into their exile work and this, combined with very great personal courage, eventually bore fruit. It was only then after many years of futile exile that the SPD began to think pragmatically about political power, how to gain it and how to keep it.

A very great deal has been written about the SPD both during the Weimar era and thereafter. Much of this, especially in the work of German scholars, has been concerned with the failure of the SPD to secure the republican democracy of Weimar, and its apparent powerlessness when confronted with the threat of Nazism.[2] It is often alleged that the SPD was not sufficiently robust in its defence of the Weimar Republic and that in some way most of its leaders were cowardly. Out of these considerations there emerged a view of the SPD which was based on its supposed shortcomings. These range from a refusal to stamp out the NSDAP, an unwillingness to take responsibility for the government of the Reich, an inability to see that the political right was more of a threat than the Communists, a weak-kneed policy towards Brüning and his successors, von Papen and von Schleicher, and finally the cowardly abdication of its political responsibilities in 1933.

Whilst the search for scapegoats for the rise of National Socialism is understandable, it is not very profitable. The objective situation that the SPD found itself in during the Weimar

---

[2] See *German Democracy and the Triumph of Hitler*, ed. E. Matthias and A. J. Nicholls (London, 1971).

years was usually very different from its interpretation by historians and others. It was not, for example, true that the SPD failed to realize the Nazis were a menace before 1933 or indeed that where it could it failed to fight them. Nor is it true that the SPD evaded the responsibilities of government. The role that the SPD played in Prussia, by far the largest *Land* of the Reich, contradicts most of the wilder allegations that have been made.[3] Whilst there is some evidence that SPD leaders may have underestimated the Nazi threat, there is far more evidence to show that where they could, they attempted to counter it.

Two of the most serious charges that are levelled against the SPD are of direct relevance to its role in exile. The first is that it refused to use force in order to defeat Hitler and his supporters at a time when only force could have prevented Nazi success. The second is that the SPD's attitude towards the Nazis in the first six months of 1933 was so equivocal as to throw the very gravest suspicions of the SPD's opposition to Hitler's policies or at any rate to some of them.

It is often argued that if the SPD had called on the pro-republican para-military formation, the *Reichsbanner*, it would have opposed with force the coup against Prussia ordered by von Papen.[4] This would have enabled the SPD to remain in power in Prussia and it would thus have become very much harder for Hitler to take power. Theo Pirker, to name but one of the SPD's critics has said that rather than standing up and fighting for the Republic it had in large part created, the SPD preferred the 'democratic answer' which was 'capitulation'.[5]

Once Hitler had assumed power it has been said that the SPD first attempted to appease him and then tried to form a 'loyal opposition' as long as Hitler agreed to remain 'constitutional'.[6] One SPD leader, Paul Löbe, even went so far as to state on 10 June 1933, only thirteen days before the SPD was declared illegal, that the party should distance itself from those leaders and members who had already gone into exile.[7] A colleague of Löbe's, Wilhelm Hoegner, a Bavarian SPD leader gives

[3] A. Glees, 'Albert C. Grzesinski and the Politics of Prussia', *English Historical Review*, LXXXIX (October 1974), 814–34.

[4] See K. Rohe, *Das Reichsbanner Schwarz-Rot-Gold* (Düsseldorf, 1966).

[5] T. Pirker, *Die SPD Nach Hitler* (München, 1965), p. 25.

[6] Lewis Edinger, *German Exile Politics* (California, 1956), p. 12.

[7] Pirker, op. cit., p. 32.

the same highly critical account of this period that many historians have produced. The SPD ought to have opposed von Papen by force, he argues, but it was incapable of action '*wir waren waschechte Parliamentarier d.h. wir redeten über die Dinge, andere aber gestalteten sie*'.[8] One of the main reasons for this failure, he suggests, was the poor quality of the SPD's leadership which consisted of bourgeois intellectuals—often of Jewish origin—and skilled workers turned trade union officials.[9]

As early as February 1933 a number of the SPD's most important leaders had decided to flee the Reich. To facilitate this they moved to Munich. They included Scheidemann, Breitscheid, Hilferding, Wels, and Hans Vogel. Hoegner personally helped many to cross into Austria, including, on 5 March 1933, Albert Grzesinski.[10] But the SPD did not only suffer from the flight of its leadership, it also embarked on an act of self-immolation, the Reich conference of the SPD which the Berlin Social Democrats convoked for 19 June 1933. Those SPD members who had fled to Prague and were now led by Vogel and Wels together, were considered by the Berlin membership to be bringing the party into disrepute. Vogel, Wels, and any other leader in Prague, it was proposed, should therefore no longer be considered members of the SPD Executive, the *Parteivorstand*, since it was 'impossible to lead the party from abroad'.[11] The Berliners argued that ever since Wels's Reichstag speech of 23 March 1933 where he had offered the Nazi regime the 'loyal opposition' of the SPD, the party had wanted to adapt to the changed situation. Indeed, on 17 May 1933 the SPD had supported a National Socialist motion that 'Germany should be treated as an equal in the international community'. To go back on this course would, they claimed, be suicidal.[12] Not a single member voted against the proposal to dismiss the Prague exiles from the Executive and a new Executive led by Löbe, Westphal, and Stelling was then elected.[13]

The folly of such action ought to have been clear to the Berlin party members. After all, Hitler had told SPD Reichstag

---

[8] W. Hoegner, *Flucht vor Hitler* (2nd ed., Munich, 1978), p. 24.
[9] Ibid., p. 22.
[10] Ibid., p. 92.
[11] Ibid., p. 299 ff.
[12] Ibid., p. 230.
[13] Ibid., p. 231.

deputies that he was not interested in SPD support.[14] Three days after the election of a new Executive the Nazis showed their contempt for the SPD by outlawing them and, within a few hours of this, arresting Löbe and over 3,000 SPD members throughout the Reich.[15]

Although it is possible to understand why the SPD should have panicked in 1933, there can be little doubt that its actions were unlikely to produce a sympathetic response amongst those whose support the SPD either had or needed to gain. Illegal action or violent revolution would, similarly, most probably have met with disastrous failure. Under the circumstances, exile was the wisest course of action. The attempt to appease Hitler and the clear division in the party made the beginning of the period of exile seem shabby and cowardly.

It should not however be assumed that this assessment of the SPD's actions in 1933 is justified. It is hardly surprising that the party should have been dispirited by 1933. It had, after all, spent fifteen years fighting the anti-republican threat. Many leaders displayed considerable courage. Vogel and Stampfer, for example, returned into the Reich from the Saar on 10 May 1933 at very great personal risk in order to try to dissuade the remnants of the SPD Reichstag party from supporting Hitler's foreign policy. Above all, however, the Nazis' use of pseudo-legality combined with quite open terrorism and violence made resistance understandably rare. It would be wholly wrong to minimize the political effect of the systematic arrest of Social Democrats after 30 January 1933, their transfer into police 'custody', and their systematic torture by the police and their SA auxiliaries in the cellars of police stations and concentration camps throughout the Reich. The SPD archives contain file upon file of documented evidence of such brutality and they can leave no room to doubt that a uniform policy against the SPD was carried out by the Nazis in order to intimidate it and to show German Social Democrats that there was no place for them in the Third Reich.

Exile, then, seemed to provide the best chance of ensuring the continued survival of the SPD. It would prove that German Socialism refused to capitulate to Hitler and it would offer at

[14] See M. Domarus, *Hitler 1932–1935*, Vol. I (München, 1965), pp. 246-7.
[15] Hoegner, op. cit., p. 233.

least some opportunity of opposing the Nazis. Although some SPD leaders may have gone into exile simply to save their own skins, most went because they felt it their duty. And, as the subsequent fates of two prominent SPD leaders, Karl Severing and Gustav Noske, show, Nazis were quite prepared to leave leaders in peace provided they went into retirement from politics. Indeed, Hitler himself once remarked that the SPD's veteran leaders inside the Reich presented no threat to the Nazis' position because unlike the Communists, they were not supported by any major foreign power.[16]

Yet the sojourn in Prague did little more than prove that the party had not gone under. Within the party leadership the endless quarrelling was to become for many the symbol of life in exile. Ollenhauer once recalled a meeting of the Executive in which one leader had attempted to strike a colleague over the head with his chair. [17] The Sopade's assessment of the effect of Nazism on the development of German political life appeared to lack any realistic comprehension. Hitler was the lackey of capitalism whose mismanagement of the economy would soon precipitate his overthrow.[18] The party paper, the *Neuer Vorwärts* repeatedly urged Social Democrats inside the Reich to 'carry out a revolutionary policy'.[19] Yet it ought to have been obvious that no one except the Nazi leadership was in a position to carry out any sort of policy at all. It is true that no SPD leader encouraged the membership actually to take up arms against the Nazis though the reasons for this are very unclear. Any organized armed resistance was out of the question, although to have encouraged individual acts of armed sabotage might have been effective. Edinger states that Breitscheid argued in 1933 that nothing could be more mistaken than to try to oppose the Nazis by force with the implication that the SPD leadership thought that to promote such a policy would simply demonstrate the SPD's impotence.[20] Yet Breitscheid may only have been urging procrastination.

In general, the exiled SPD's contribution to Social Democratic ideology was unimpressive. Hitler was seen as simply a

[16] H. Trevor-Roper, *Hitler's Table Talk* (London, 1953), p. 21.
[17] SPD, *Mappe* 79.
[18] Edinger, op. cit., p. 36.
[19] Ibid., p. 45.
[20] Ibid., p. 44.

symptom of the death throes of capitalism and this clouded thinking on many policies including those concerning the possibility and nature of effective resistance to Nazism. The theoretical baggage that the exiles carried with them to Paris and later to London was irrelevant to the real demands of exile policy-making.

The notion that Nazism was but another face of monopoly capitalism, Kautsky's view that Hitler would never achieve the 'totalitarianism of Fascism' or that he would fade away through his inability to score any successes, were theories as redundant by 1938 as the hope that the Nazis would succumb to strikes and ultimately a revolution.[21] Indeed, it was precisely the uselessness of such ideas which may have prompted the exiled leaders in London to concentrate on practical politics. They were not distinguished by their gifts as political theorists, and although they brought with them attitudes that could not be discarded, these were not permitted to play any significant role in determining action.

It is plain that no clear messages came out of Prague. Nothing was said or done to indicate to the SPD's membership inside the Reich that the real constraints on it were understood or that the leadership was pursuing new alternative policies which might have helped in the fight against Hitler. Rather, the main purpose of the Sopade Executive in Prague was to maintain links with a fast disappearing party organization and to glean as much intelligence about what was going on inside Germany as it could. The usefulness of such work was, however, counterbalanced by the recriminations that were going on inside the Executive. In August 1933 Aufhäuser and Böchel, two prominent left-wingers, were co-opted onto the Executive in order to answer the criticisms of many younger SPD members that the old leadership was discredited and failing to produce new solutions to the problems of the SPD.[22] The former argued that the new Germany could only be defeated by élitist socialist vanguards and that full democracy had proved wrong for Germany.[23] It was the popularity of such radical ideas amongst the exiles that led to the formation of the *Neubeginnen* ('new beginning') group of about 300 SPD members who continued

[21] Ibid., pp. 9, 16, 20, 110, 118.
[22] Ibid., p. 44.   [23] Ibid., p. 106.

throughout the exile period to act as an important ginger group. Similarly, in October 1933 Curt Geyer, who had until then been the underground leader of the Berlin SPD, joined the Executive, which now consisted of nine members, because he had had first-hand experience of the struggle against the Nazis. It seems that the Executive was determined to answer the rank and file call for 'a radicalisation of its position in the battle against Hitler'.[24]

One result of this radicalization was the Prague manifesto of 1934. It stressed the Marxist traditions of the SPD, its belief in the existence of class struggle, and the fact that it had been a crisis of capitalism that had brought Hitler to power. It urged Germans to accept the Sopade's offer of 'revolutionary socialism and its revolutionary leaders'. At the same time, however, the Executive refused to abandon its hope that a variety of political and economic factors rather than armed force would bring about Hitler's downfall. These included an internal struggle within the Nazi party for which the Röhm affair seemed to offer some evidence and a *putsch* by the German army, a theory especially favoured by Stampfer who may have had some contact with the army.[25]

Yet it seems very hard to believe that the declarations of the Sopade possessed a good chance of appealing to the German people inside the Reich. What was needed was some sort of positive political action. One opportunity for this appeared to present itself after the summer of 1935 when the German Communist party (KPD) began to make overtures to the SPD. The Comintern which had until then branded Social Democrats as 'Social Fascists' decided to alter this assessment and it encouraged the KPD leaders to attempt the formation of a popular front together with the SPD against the Nazis.

From the SPD's point of view a coalition with the KPD seemed to offer some attractions. For a start, it would be a chance of overcoming the rift in the German Labour movement which, since 1917, had done so much to assist the enemies of Socialism. Second, it would give the SPD much needed allies and, perhaps, a friendly nation which might offer support. Finally, it was a chance for the Sopade actually to do something of political significance. Some SPD leaders thought the

---

[24] Ibid., p. 114.   [25] Ibid., pp. 138-9.

party should pursue the Communist offer. Aufhäuser was their leader, arguing that the time was now ripe for the formation of a 'united revolutionary Socialist party' whose 'natural ally would be the USSR'. But a majority of the Executive was very strongly opposed to the whole idea. A crisis loomed and Aufhäuser and Böchel were forced to resign from the Executive.[26] The most adamant opponents of any deal were Vogel and Stampfer who stated that the KPD did not really represent the interests of the workers in Germany but those of the USSR.[27]

By the autumn of 1937, however, the danger of a serious schism in the SPD over this question had receded. This was perhaps because of the troubles that Léon Blum's popular front arrangement was beginning to cause. At any rate, Böchel rejoined the Executive and even the younger SPD members now refused to have any dealings with the KPD. The *Neubeginnen* became the focus of their interest and it was dominated increasingly by men like von Knoeringen, Schoettle, Karl Frank alias Willi Müller and Paul Hagen, and Richard Loewenthal who all achieved prominence in the years to come. The Sopade Executive itself began to be less radical, extolling the Weimar virtues of liberal democracy rather than revolutionary socialism, something which began to be explicit in the manifesto of January 1936.[28]

After 1937 it seems that the exiled SPD did little other than compile warnings about Hitler which it sent around the world. Erich Rinner, who edited these *Deutschlandberichte*, sent these reports to statesmen hoping they would encourage opposition to the Nazis and gain acceptance for the SPD interpretation of affairs.[29] Apart from this there were few things to occupy the Social Democrats. The Executive was not only short of funds but the last vestiges of party organization in the Reich had disappeared.[30] The leaders of the SPD had become 'generals without an army' who were not worth taking seriously.[31] In addition the SPD's often repeated strong opposition to the

[26] Ibid., p. 155.
[27] Ibid., p. 157.
[28] Ibid., pp. 168, 176.
[29] Ibid., p. 190.
[30] Ibid., p. 190.
[31] Ibid., p. 194.

policy of appeasement made it an uncomfortable comrade for some European socialist parties. Hilferding, for example, had written in 1935 that 'the pacifism of the British Labour party had constituted the strongest support for Hitler'.[32]

But the gravest problem that the SPD leadership had to face was a Nazi invasion of Czechoslovakia. In 1933 Beneš had declared that his country was proud to be a haven for exiles (he had, of course, been one himself) and Wenzel Jaksch, the Sudeten German Social Democratic party leader, gave parliamentary support to the Beneš coalition.[33] Yet the genuine goodwill that was shown to the SPD was, by 1937, being undermined by Nazi threats against Czechoslovakia because it harboured 'the enemies of the Reich'. Indeed, on 21 May 1935 Hitler had declared in the Reichstag that 'outside our border there is an army of exiles working against Germany'.[34] By December 1937 Beneš felt forced to tell the SPD leadership that they would have to cease their anti-Nazi activities in public within a few weeks.[35] This caused the Sopade Executive to contact Léon Blum to see whether they would be allowed to work in Paris and when an affirmative reply was received, the Executive left for Paris. This did not bring a halt to Hitler's diatribes, however, nor to his giving Beneš's support of 'anti-German' groups as a pretext.

But Paris was not a new beginning. In the first place, by the time the SPD set up headquarters at 30 Rue des Écoles, Blum had fallen and the new French leader Daladier offered nothing like the welcome the SPD had expected. Secondly, the Executive seemed reluctant to continue its work and took some time to contact comrades in France. The most significant actions taken by it in Paris were, first, the sending of a mission to London in the winter of 1939 and second, the sending of Stampfer to America in January and February 1939 to make contact with the American Federation of Labor and the Jewish Labor committee in order to gain both visas and money if it became necessary for the SPD members and leaders to flee to the USA. In March 1939, with Stampfer's support the German

[32] Ibid., p. 197.
[33] Ibid., p. 281.
[34] K. Grossmann, *Emigration* (Frankfurt, 1969), p. 69.
[35] Edinger, op. cit., p. 203.

Labor Delegation was set up in New York under the chairmanship of the robust Albert Grzesinski.[36] In 1941 it became the Association of Free Germans Inc.

An examination of the papers of the SPD Executive in Paris indicates the extent of the disintegration of German Social Democracy after 1938. Despite weekly meetings, the only serious work that was done concerned the provision of aid for refugees.[37] Social Democratic refugees' aid centres, as they were called, were set up in London under Wilhelm Sander whose name was symbolically shortened to Wisa and in Stockholm where Emil Stahl was made 'party trustee'. For both Britain and Sweden had become important centres for refugees. When on 20 June 1939 the Executive decided to compile a report on its other activities the results were modest indeed. The SPD had two aims, it was stated, the first being to monitor internal developments inside the Reich and second to ask the German workers to 'work more slowly' but not to encourage passive resistance to the Nazis or sabotage 'because the Social Democratic worker owes his good standing in the factory to the hard work that he does'.[38]

It is hardly surprising that the SPD leadership in Paris was attacked for its alleged inactivity. A former Executive member Paul Hertz, who had been expelled in 1938, called a meeting of all Socialist exiles in Paris in August 1939. He argued that the idleness of the Executive was seriously undermining the coherence of the exile fight against Hitler. The SPD leadership, he claimed, lived a bogus existence, a *Schattendasein*, doing nothing except office work of the most routine kind, a bureaucratic *raison d'être*. The SPD Executive felt forced to counter this potentially damaging charge of indolence, but with no

[36] Grzesinski who had been a Prussian Minister of the Interior had found work in a jewellery factory in New Jersey. See Glees, op. cit.

[37] SPD *Mappe* 11.

[38] Ibid. This slogan was used last in 1936 and with no apparent success because, the SPD leaders claimed, 'of the international situation'. But within a month of the re-adoption of the slogan it was dropped yet again on 5 July. Fritz Tarnow's arguments against the slogan were adopted. It was thought that it could easily lead to counter-propaganda viz. that the SPD did not really have the best interests of the German workers at heart and did not correspond to the wishes of German workers to work more, work the Nazis could offer because of their preparations for war. This decision to drop the slogan pinpoints an important dilemma for the SPD at this time not least because the 5 July meeting was specifically called to decide on SPD policy should war break out.

great conviction. It had, it said, organized every SPD member who could prove his membership before the cut-off date of 1933 and all 'comrades met as often as possible'.[39]

There were problems, the Executive agreed, but these had been caused by the fact that there had already been a national SPD group, or *Landesgruppe*, in France and so there had been some confusion amongst the membership when the Executive arrived. Organized Labour in every nation where SPD members had grouped themselves, officially recognized the SPD Executive in Paris and it was, in addition, 'in continual contact with the British Labour party who also helps our refugee aid committee in London'. All this had been achieved despite there being only six elected Executive members as opposed to twenty in the past and the fact that half the 1933 Executive were either still inside the Reich or overseas. Finally, the SPD continued to bring out the *Neuer Vorwärts* and the *Deutschlandberichte* which it 'was generally agreed provided the best information about Germany in existence'.

The Executive was, however, not unmoved by Hertz's criticism because it decided to revitalize the Executive meetings by inviting Wenzel Jaksch the leader of the Sudeten German Social Democrats and Erich Brost to attend.[40] When the Nazi-Soviet non-aggression pact was signed in the summer of 1939, the Executive decided to make an official statement. It said that peace had been betrayed and that the Soviet Union was now 'playing Hitler's game'. It argued that 'in the struggle between the warlike and expansionary policies of German Fascism and the peace front of the European democracies, the USSR has come out on the side of Fascism'. Stalin, it pointed out, would bear the full responsibility for the increased danger of war and it was Stalin who was blamed for the blow which had been struck against the German opposition. Nevertheless, the SPD would continue to fight for the overthrow of the 'Hitler-dictatorship', it promised, and it asked the 'world to recall that with the support of those inside Germany, the

[39] Ibid.; see W. Röder, *Die deutschen Sozialistischen Exilgruppen in Grossbritannien* (Hanover, 1969), p. 28.

[40] See below, p. 237. Erich Brost, an SPD leader from Danzig, had wide experience in journalism and after the war went on to edit the *Westdeutsche Allgemeine Zeitung*, the paper with one of the largest circulations in West Germany. For Jaksch see below, p. 45.

SPD had always refused to collaborate with the German Communists'.

When a little while later war was actually declared the SPD decided to issue another statement, this time to the German people.[41] It would, however, be difficult to see this declaration as any more convincing than the one above. Although its aim was to prove to the external world that the SPD still possessed a domestic German following and to prove to Germans inside the Reich that the SPD was listened to in the councils of Europe, the self-justificatory tone of the document was more likely to achieve the opposite effect. The Executive, in appealing 'to the German people and the whole world', reminded this not small audience that it was the 'last body to have been elected by the Social Democratic mass organization in Germany' and was therefore able to 'speak for the party, and for those parts of the German population who oppose war and dictatorship'. Their common goal was the 'overthrow of Hitler and the new German militarists who are one and the same'. The SPD offered to fight together with the German people in order to 'overthrow Hitler and to restore freedom and create a peace which makes good all acts of violence and puts an end to all totalitarian systems and dictatorships'. It was, the statement continued, 'retrogressive and superstitious to believe that the welfare of the German people could depend on territorial aggrandisement'. Rather, the 'true historical command' was the peaceful reconstruction of Europe. Finally, the SPD would now become 'an allied force fighting for freedom alongside all Hitler's opponents, working during the war on behalf of European culture'.

If such a statement was to gain acceptance and credibility it had to convince those with authority that the SPD was in a position to be listened to inside the Reich and it had to convince those inside the Reich that what was suggested was practicable. The SPD Executive's intention in issuing this declaration is plain. It is far harder to see how it could ever have been successful, for even if Germans had been able to read the statement or hear it, they were not being given any concrete advice about how to overthrow Hitler and the new militarists or how to prevent war. As the confusion over the slogan 'work more

[41] SPD *Mappe* 11, *An das Deutsche Volk.*

slowly' had indicated, it was very difficult indeed for the SPD to discover an appropriate form of discourse with Germans inside Germany. As long as the tide seemed to be flowing in the Nazis' favour, the SPD's problems could only increase. Although bad news and military defeats might not necessarily enhance the SPD's reputation, its fortune could not be improved without German defeat. The international community would only pay serious attention to the SPD when it could demonstrate that what it said was listened to and acted upon. If the SPD no longer spoke for the interests of German workers or they no longer recognized it, its own claim to further political existence was called into grave doubt.

It was not until Hitler was almost at the gates of France that steps were being taken to prevent the idea of German Social Democracy from disappearing altogether. In 1940 the party issued three declarations which showed a realism and an urgency previously absent.[42] In January the German people were warned to avoid any complicity in the Nazi atrocities against the Polish people. The SPD told of the mass murders that were taking place which were 'part of a systematic attempt to annihilate an entire nation' and it warned that unless the German people took action 'to destroy Hitler, the victims of Nazi policy will blame the entire German people'. Similarly on 20 April, on the occasion of Hitler's attack on Scandinavia, the SPD demanded from the German people 'disobedience, resistance and rebellious opposition against the band of criminals who now hold the reins of government'. And in May 1940 a message equally harsh and bitter was sent to Germany by the SPD. The awful things that were about to happen were the fault of Hitler

the criminal to whose rule you have submitted yourselves for seven long years. If you share Hitler's guilt for his crimes, Germany's future will look bleak. By your silence, your tolerance or even your apparent approval, the world gains the impression that Hitler's will is the will of Germany.

It concluded with an appeal for opposition, 'take every means of making opposition to this criminal war . . . Hitler must die so that Germany may live.'

There can be little doubt that all this is evidence of the first stirrings of a new approach by the SPD leadership. It seems

[42] SPD *Mappe* 11, declarations of 29 January 1940, 20 April, and 10 May 1940.

that they believed the demands of exile politics required them to move in a different direction. Vague talk of revolution and Marxist dreams about the future were giving way to stern warnings and an almost frantic plea to the German people to realize just what Hitler was doing to them and to the rest of the world. Beneath these stirrings, implicit in them, was probably a change in the SPD Executive's perception of its role and of its relationship to the people it claimed to represent. It was not that the SPD's leaders believed the German people had let them down but rather that one means of communication between leaders and led was breaking down and another needed to be found. An important part of this process, it was thought, lay in gaining the support of the international socialist community and one advantage of talking harshly to the German people was that it must have pleased those whose support the SPD wanted. Above all, the SPD was now very firmly supporting those at war with the Third Reich.

In March 1940 Vogel developed the SPD's new attitude towards its task at a meeting of the Socialist International in Paris.[43] The SPD was now ready and willing to stand alongside the Western democracies in their fight against Hitler. They were 'the other Germany', he asserted, and an ally. At the same time he expressed the hope that Germany would be freed by an internal rising rather than by 'foreign bayonets' and that the Socialist International would do everything it could to promote internal opposition.

The SPD leadership was, then, now prepared to put itself into the camp of those who had declared war on Hitler. This was a situation without precedent for German Social Democracy. It had experienced exile before, it had experienced a conflict of conscience before of which Bebel's opposition to Bismarck's annexation of Alsace Lorraine is only one example, and it had experienced the 1914–18 war. But it had never actually offered aid to Germany's enemies. The courage of the Executive should not be minimized, for its move was highly dangerous for the future prospects of the party. The Nazis could easily claim that Social Democrats were traitors, that their demand for the overthrow of Hitler was high treason, that Social Democrats had once again been exposed as *vaterlandslose*

---

[43] Ibid.

Gesellen. In addition, the SPD leadership no longer based its demand for international recognition on the fact it was the voice of the German workers but claimed to be the voice of all those who opposed Hitler, even if they had so far remained quiescent.

The position of the SPD in Paris worsened in keeping with the international situation. In May 1940 Ollenhauer, and one of his sons, Peter, were interned for a few weeks, but later released.[44] But in June 1940 the French government capitulated and for the first time since 1933 the SPD leadership was faced with the prospect of physical extinction. In the June armistice which Hitler had dictated to the French authorities at Compiègne, they agreed to hand over any ex-German nationals whom the German government asked for. Most SPD leaders decided to flee into the unoccupied zone of France and Marseilles. Here they hoped to receive visas to enable them to leave for the USA or Britain although there was a rush for berths to any part of the world that would take the refugees in and exiles left for South America, Hong Kong, and Macao.

In a desperate bid to leave Paris before the German army arrived, the SPD headquarters at the Rue des Écoles was locked up, the party funds deposited in a strongbox in a Paris bank and Vogel, Ollenhauer, Hilferding, Breitscheid, Heine, and Geyer left for the south. Vogel was now chairman of the Executive for on 16 September 1939 Otto Wels had died. Another senior member, Siegfried Crummenerl, died in May 1940. Ollenhauer and Vogel decided not to wait in Marseilles where they were daily threatened with possible arrest by the Vichy policy but to try to reach Spain or Portugal through the Pyrenees, taking their families with them.[45] Heine and Geyer, who had been co-opted onto the Executive in 1938, stayed in Marseilles to co-ordinate the ever-worsening refugee situation. Heine's dispatches and pleas to SPD members throughout the world make a gripping account of human beings trying frantically to escape an increasing threat of destruction.

There were roughly 140 SPD members in the unoccupied

[44] SPD *Mappe* 79. Léon Blum secured their release. See Röder, op. cit. (1969*a*), p. 27.
[45] SPD *Mappe* 79.

zone who together with their families constituted about 600 persons.[46] There was a grave shortage of money and valid visas. The asking price for a $100 passage was anything up to ten times that. Heine made frantic appeals but support was slow in coming and the French authorities interned more and more refugees in the notorious camps of Gurs and Argeles.[47] By December 1940 there were over 15,000 inmates in the former and 10,000 in the latter. Many refugees gave up all hope of escape and either died in the freezing conditions (the camps had no heating) or killed themselves. On Christmas Day 1940 Heine wrote that twelve telegrams and one hundred letters had produced neither the visas nor the money he needed so badly.

Some aid was forthcoming but it was barely sufficient for more than a handful of refugees. The Jewish Labor committee sent Heine $1,250 or over FF 22.000.000 for the refugees[48] and Hilferding received a personal gift of $100 from the ex-Chancellor of the Reich Brüning.[49] For those rounded up by the French police the fate of Hilferding, which was also shared by Breitscheid, served as a chilling foretaste of what they could expect. On 6 February 1941 the American government finally agreed to offer them visas but the Vichy police decided not to recognize them and they were revoked four days later. In the early morning hours of 13 February French police arrested them and their wives and they were driven back to Paris. The men were then handed over to the Gestapo. Their subsequent history is uncertain. The SPD leadership was given to understand that they were taken to Berlin on 10 March 1941 where they were interrogated before being deported to their deaths.[50]

The physical disintegration of the SPD Executive was matched by the political disintegration of the SPD as a party. By Christmas 1940 it seems almost impossible to speak of the SPD as a political party in any meaningful sense. Its external activities had dwindled and its internal activities were almost

[46] SPD *Mappe* 51, see letter to Nielson from Heine, 30 May 1941.

[47] See below p. 71.

[48] H. E. Tutas, *Nationalsozialismus und Exil* (München, 1975), suggests that Hilferding committed suicide whilst in prison in Paris and Breitscheid died in Buchenwald.

SPD *Mappe* 51, ex-President Beneš wrote to Vogel of his grief at the fate of these two men.

[49] Ibid.

[50] SPD *Mappe* 18.

wholly concerned with refugee work. There is no evidence that
any SPD leader thought about the future: understandably perhaps since the present was sufficiently taxing. It seemed that
the journey out of Germany and into exile, which it had been
hoped would preserve the party and uphold its historical traditions, had by the first winter of the European war failed in
its purpose.

The objective situation, then, appeared to suggest that, however important the SPD leadership's new approach to exile
politics might be, physical destruction at the hands of the
Nazis would probably be the party's fate. But it was precisely
at this desperate moment that the dispatch of a deputation to
Britain, one year before, bore a vital fruit. The talks that this
deputation had conducted with the British Labour party and
the British Foreign Office were to lead directly to the SPD's
continued survival as a major political party.

II

First Contact with England

THE exiled SPD leadership had maintained links with the British Labour movement since the summer of 1933.[1] At that time the Secretary of the International Subcommittee of the National Executive of the Labour party, William Gillies, had compiled a list of prominent SPD members in London which had been requested by Otto Wels.[2] These included Viktor Schiff who became an influential journalist on the *Daily Herald*, Hermann Badt who had been an aide to Albert Grzesinski in the Prussian Ministry of the Interior, Julius Braunthal, the lawyer Otto Kahn-Freund, and the ex-*Reichsbanner* leader Karl Hoeltermann. Gillies was a figure of very great importance in the foreign policy-making of the Labour party. And as such he was destined to play a major role in the affairs of the SPD in exile. A Scot, he had come to London in 1917 where he had created an international research department for the Labour party becoming the 'international Secretary of the Labour Party' in 1919. He held this post until 1946 when he was replaced by Denis Healey.[3]

Since 1938 Wilhelm Sander, or Wisa, had worked on refugee matters for the Executive from an office in Mill Hill in London and from that address he also published the *Sozialistische Mitteilungen*, a newsletter 'for German socialists in England fighting Hitlerism and dictatorship of any kind'.[4] Sander's visa for the UK had been personally secured by Gillies. Hans Gottfurcht, an ex-trade union official from Berlin whose

[1] In 1929 Fritz Heine had observed the election campaign for the SPD Executive.
[2] SPD *Mappe* 68, report dated 8 June 1933.
[3] See below p. 141 and *The Bystander*, 3 July 1940. Also B. Reed and G. Williams, *Denis Healey and the Politics of Power* (London, 1971), p. 52.
[4] SPD *Mappe* 11.

interest in these matters caused him to become the SPD leadership's trade union advisor, had also settled in London at about this time. In July 1939 Gillies sent the SPD leadership in Paris £150 to pay for the printing of 20,000 leaflets which were to be distributed inside Germany and he also donated RM 15,000 towards the cost of their being smuggled into the Reich by couriers.[5] The leaflet described the attitude of the SPD towards the coming war and it is interesting that the Labour party believed it was important that the SPD's view should be brought to the attention of the German people.

The co-operation between British and German Labour movements was conducted, at this stage, in a spirit of solidarity in the struggle against Fascism and also in a spirit of equality. There is no evidence that the SPD leaders were either looked down upon because they were German or that they were in any way blamed for the excesses of the Third Reich. The SPD leaders took this co-operation extremely seriously, and they too saw no reason why they should not treat the Labour leaders as equal partners. In August 1939, for example, they wrote to Gillies that despite their 'pleasure at hearing the appeal of the Labour council to the German people' they were highly critical that 'some hope of a compromise peace appeared to be held out which would lead only to the consolidation of Hitler's grip on Germany'.[6]

As the international situation deteriorated after September 1939, the significance of this co-operation increased. There was of course the major risk that Hitler might decide to invade France, in which case the future of the SPD in exile would have to be considered. Ought the SPD leadership to give up the fight? Ought its leaders to flee to America? Or ought they to transfer their headquarters to London? In a letter to Vogel of 5 September 1939 marked 'secret' Gillies wrote about these matters. He made it quite plain that the British Labour movement would be delighted if the SPD were to move its official seat to England. Now, according to Gillies, was the time to think about 'co-operating with us in this country' and he sent £100 as a token of the Labour party's goodwill towards the SPD.[7]

[5] Ibid.   [6] Ibid.
[7] SPD *Mappe* 68.

Gillies's helpful overtures were taken up by the SPD leaders in the winter of 1939. In December Vogel and Ollenhauer decided to accept the invitation of the British Labour party to come to London for talks. There can be little doubt that this visit was to add further impetus to the co-operation that had already been going on and, more than this, to influence the SPD exile so much that it decided to move to London. Not only were German Socialists treated with respect and sympathy, they were also given every indication that, were they to come to London, they would be treated as equal partners in the developing fight against Hitler and the Third Reich. A number of the important leaders of British Labour met Vogel and Ollenhauer, and Gillies also arranged for them to be received by the Foreign Office.

The two SPD leaders met Attlee, Dalton, Noel-Baker, and of course Gillies himself, as well as Bevin and Greenwood who represented the Trade Union movement. At this meeting, Vogel and Ollenhauer pressed their case for receiving financial support on a regular basis from the Labour party.[8] This was agreed. In return the SPD would do everything in its power to assist the British in their propaganda war against Germany. In addition Vogel and Ollenhauer agreed to ask SPD agents inside the Reich to monitor British broadcasts to Germany and to convey their findings back to Britain through the good offices of Gillies. The participation of the SPD in a 'shadow government', or government in exile, was also discussed. It was suggested that Breitscheid might be a suitable nominee from the British point of view but 'for the time being' the SPD rejected the idea.

There seems little doubt that the two SPD leaders were greatly impressed by the comradeship shown to them by the British Labour party. They were possibly even more impressed by the real opportunities for political work that their transfer to London seemed bound to bring about. Not only did the Labour party express willingness to offer financial support, which was to become even more vital after Hitler's invasion of Scandinavia cut off many German Social Democrats and prevented them from supporting the party or buying its

[8] Labour Party Archives (LPA) Middleton papers Box 8, see Vogel's letter to Gillies, 8 January 1940.

publications, but it also seemed ready to grant the SPD access to its leaders and policy-makers.[9]

There was also a geographical reason for choosing exile in Britain rather than following the example of a number of leading German Social Democrats and going to the USA.[10] The physical proximity to Germany combined with the fact that America was as yet still neutral helped the SPD to maintain an identity. Those who did go to the United States, on the other hand, do not appear to have achieved the sort of political success that their English colleagues gained. Many exiles in the USA began to dissolve their ties with Germany on arrival. Wilhelm Sollmann changed his first name to William, Karl Frank the vivacious *Neu Beginnen* leader gave up politics altogether in order to become a psychiatrist, and even Albert Grzesinski who for the duration of the war continued to do part-time political work, decided in 1945 when actually confronted with the prospect of returning to Germany, to take up American citizenship and not leave his new home. Those who had gone to America and nonetheless hoped to return, like Stampfer, found many obstacles in their path, as did Heinrich Brüning when he considered a return to political life in Bonn.

The USA had, of course, a tradition of integrating exiles which Britain did not possess. Foreigners could be accepted in America with greater ease than in Britain. SPD leaders in London were socially completely isolated. Neither Vogel nor his wife ever learnt to speak English and they were never encouraged to feel at home in Britain by the sort of facilities for German speaking refugees that America offered. Yet almost all of the Britons they encountered, including Gillies himself, could speak German.[11] Thus it is not surprising to realize that political exile in America did not produce the sort of political continuity that British exile offered.[12] Although Grzesinski and Katz formed political organizations and had access to two officials in the State Department, and although a number of American institutions offered posts to Germans like Sollmann, Arnold Brecht, and indeed Brüning, who taught at Harvard, the

[9] SPD *Mappe* 11.

[10] See J. Radkau, *Die deutsche Emigration in den USA* (Düsseldorf, 1971).

[11] Interview with Rt. Hon. Lord Noel-Baker, 27 Nov. 1977 and Fritz Heine 21 Aug. 1975.

[12] Radkau, op. cit., p. 19.

absence of a strong and determined Labour party in America which might have had an interest in promoting exile fortunes meant that such opportunities simply integrated exiles more deeply in American life.[13]

The optimism that Vogel and Ollenhauer felt about Britain can only have been reinforced by the meeting they had with officials at the Foreign Office. The British Labour party set very great store by gaining Foreign Office acceptance of the SPD as a legitimate representative of the German opposition to Hitler. British Socialists did not like the fact that the only refugees from Germany to whom the Foreign Office appeared to listen were right-wing.[14] It was supposed, perhaps not unreasonably, that it would have an obvious preference for conservatives and would be inclined to ignore SPD views. Consequently, Gillies arranged on 17 December 1939 with H. M. Gladwyn Jebb the Private Secretary to the Permanent Under Secretary, Sir Alec Cadogan, that Vogel and Ollenhauer should be received 'by someone in authority'.[15]

In a 'very confidential' letter to Sir Alec Cadogan, Jebb informed his chief of all that had gone on. Gillies had come to see him 'for the first time in years' in order to ensure that the SPD leaders were taken seriously. Vogel 'was a Social Democratic leader who used to edit the Sopade *Berichte*, that green anti-Nazi pamphlet published secretly in Prague'. Gillies, according to Jebb, was extremely cross because 'those émigrés who have succeeded in establishing contact with the powers that be, and thereby exercise a certain influence on British policy, are recruited almost exclusively from the right'. Gillies specifically mentioned the case of Rauschning, 'a German nationalist, even if he is able, straightforward and honest'. Gillies claimed that although Rauschning's aim was the overthrow of the Nazi regime, all he wanted to establish 'in its place was a regime directed by benevolent generals'.

Naturally their principle object would be to preserve as much of the present *grossdeutsches* Reich as they can and equally naturally they will restore a system of private enterprise. In other words they would call back

[13] Radkau, op. cit., p. 183. Grzesinski was appointed official informant on German political exiles for the OSS. See also W. Link 'German political refugees—the US during World War 2' in Matthais and Nicholls (eds.), op. cit.

[14] See below p. 41.

[15] FO 371.24420 c 897, 17 December 1939.

into power their old allies, the Junkers, the *Grossindustriellen* and many representatives of the 1914 Germany.

Gillies made specific mention of Stinnes and Thyssen and his argument was thus made plain. Unless socialist policies were adopted, Jebb was told, and unless socialist solutions were used there would be a resurgence of those forces which had, in one generation, produced two world wars. Gillies clearly believed that the only sure way of achieving real peace in Europe lay in enabling the SPD to shape the new Germany.

Yet Jebb did not believe that Gillies was simply being what Jebb called 'ideological'. He thought Gillies was really afraid that

all the influence of the right-wing émigrés would be directed towards inducing the British government to conclude some patched-up peace. If the war did not go very well for us during the first six months or year, the danger of their succeeding was substantial.

Gillies thought that France would prove to be weak internally and he had been most upset at 'hearing Captain Margesson [a Conservative whip] declare that if only Hitler could be got rid of, we could easily fix up a peace with Goering and then all have a go at the Bolsheviks'. Gillies and British Socialists thought such

a view must be criminal lunacy. Unless the Germans are defeated we should have the same business all over again in a few years time, with this difference only, that we should as a nation be in a far inferior position to resist.

Gillies nevertheless thought that Hitler would be defeated. After it he suspected that anarchy and confusion would reign in 'that unhappy land'. But what was vital was that political power should not go to those reactionary forces which had produced the war. His conclusion was significant both because it summarized the view of the leadership of the British Labour party and because it illustrated the key role the SPD believed it would be able to play—not only during the war but after it as well:

the persons *we* should wish to see restore order should not be the militarists and the bosses but the Social Democrats. It would be far more easy to associate Germany, or Germanies, run by them with a western union than to do the same with a state imbued with the ideals of *Preussentum*.

This, Gillies argued, was the reason that the Foreign Office should receive the two exiled leaders of the SPD, Vogel and Ollenhauer.

The attitude of the British Labour party towards the SPD at this stage was plain. The SPD leaders could not merely help with the war effort, with policy-making, and propaganda. Also, they could expect to be given prominence by the British in post-war Germany. Indeed, in the process of rebuilding and purging Germany, Social Democrats could count on British support and expect to be treated as comrades and allies. Hence Vogel and Ollenhauer had the right to believe that the SPD in exile had come far since 1933, and had changed very much since the intrigues of Prague and the flight from Czechoslovakia. Its political future after Hitler seemed assured.

Gillies's interview with Jebb ended with their considering what seems now to have been but a minor point. Oddly enough, however, it concealed an area of savage future conflict, one which was, in part, to undo all that was being planned. Gillies asked Jebb why the British authorities continued 'to indulge in the dream that Germany would, in the absence of some major defeat, settle down into being an unaggressive member of the European fraternity'. Why, he queried, 'do we still make such pathetic efforts to distinguish between good and bad Germans? Surely anyone who knows anything of European history and politics must be aware that this distinction is a sheer impossibility'.

Gillies had taken great pains to assist the SPD in exile. He had gone to great lengths to arrange useful talks for its leadership and had, in Jebb's words, for the first time in years, decided to come to the Foreign Office to make representations on its behalf. Yet this small point indicates quite clearly that he did not think the SPD's claim to political importance had anything to do with the extent of support for it inside Germany. He did not believe it was the political expression of a majority of democratic Germans. It was simply to be a tool, an agent of those in Britain who wished to change the apparent course of German history. Slight though his comment was, it nevertheless gives a hint of Gillies's conception of the SPD's future role, and it was a role the SPD proved quite unable to assume. At the same time, however, it was precisely because the Foreign Office continued to make this distinction between 'good' and 'bad' Germans that Gillies's suggestion was taken up and Vogel and Ollenhauer were received.[16]

A visit by the SPD leadership in exile to the Foreign Office was a matter of considerable significance. At the same time, however, this incident illustrates the very poor quality of the Foreign Office's knowledge of German Social Democracy. Vogel and Ollenhauer can have realized only very little of this. From their point of view it was important to have been interviewed. But, as is discussed in greater detail below, the British ignorance of the SPD is somewhat surprising.

An official noted his department's 'enormous interest' in meeting the SPD leaders although he was not certain precisely how the SPD could be utilized since 'their leaders' quarrel with Hitler is that he sits on the chair they want'.[17] The official continued unabashed that Vogel cannot have been the editor of the Sopade *Berichte* 'though if he did he must have done so under Otto Strasser's direction for he always claimed to be editor in chief'. Anyone with a minimum of understanding of German exile politics would have realized how foolish these remarks actually were, for Strasser, an ex-Nazi leader, had his own pseudo-fascist exile organization and no Social Democrat would have ever worked with him.[18]

Yet this official did concede that an exile group which wanted both a war against Hitler and a complete reorganization of German political life was worth talking to. So far there had only been 'two German refugees, Treviranus and Spiecker, who are not Jewish and yet say openly that the war must be finished by the sword if there is to be any peace in Europe'. Thus it was decided that Gladwyn Jebb should receive Gillies, Vogel, and Ollenhauer, whose name Jebb told Cadogan he had 'unfortunately been unable to catch'.[19] Jebb thought the exiled Social Democrats 'quite sensible although they did not have anything particularly novel to say'. Gillies restated his belief that they ought to be used as an 'antidote to the right-wing informers' and Vogel asked the British to give the German people the straight facts about the international situation and

[16] See below, p. 57.
[17] FO 371.24420 c 897, 17 Dec. 1939.
[18] This was called the Black Front. See O. Strasser, *Germany Tomorrow* (London, 1940).
[19] FO 371.24420 c 897, 17 Dec. 1939.

'provide more news of speeches and statements of English Labour leaders'. At the same time Vogel noted he did not share 'Mr. Gillies's view of Germany and the Germans'.

Jebb suggested that Vogel and Ollenhauer make a start on their political activities by going to see Sir Campbell Stuart. Sir Campbell Stuart, a Canadian by origin, had conducted British propaganda activities during the First World War and had also co-operated with Lord Northcliffe in America.[20] He was a highly influential figure who had been brought back into these matters at the time of the Czech crisis of 1938. He was given an office at Crewe House. During the 'phoney' war he had done 'practically nothing' and during 1939 he and his outfit were moved to Woburn Abbey.[21] Gillies, however, on hearing Campbell Stuart's name, expressed the 'utmost indignation'. His organization was 'in touch only with right wing people and he, Gillies, felt up against a brick wall when dealing with this department'. His advice on propaganda was never sought and, so Jebb reported to Cadogan, he had at times felt like 'abandoning all interest in German propaganda because of this'.

More than anything else Gillies expressed dismay that Jebb had wanted Vogel and Ollenhauer to see Campbell Stuart without his being present at their meeting. He insisted that all communications between the Social Democrats and Campbell Stuart were to pass through him and Transport House. It is interesting to note that William Strang, who read Jebb's report, minuted that 'Gillies needs very careful handling... otherwise there will be considerable trouble with Transport House'. Gillies was clearly regarded as a powerful political figure and, therefore a most useful ally for German Social Democrats.

What passed at the meetings with Sir Campbell Stuart's department was recorded by Sir Campbell Stuart himself on 15 January 1940.[22] Gilles started off 'in truculent vein in order to impress the exiles'. He complained about the inadequacies of British leafleting of Germany and claimed that it was controlled 'by a German clique which was reactionary and anxious

[20] See Sir Campbell Stuart, *The Secrets of Crewe House* (London, 1920) and his memoirs *Opportunity Knocks Once* (London, 1952); FO 371. 22898 c 12865, 24 Aug. 1939.

[21] Sir J. Wheeler-Bennett, *Special Relationships* (London, 1975), p. 152.

[22] FO 371.22898 c 12865 and 24420 c 897.

to see the Junker class of pre-1914 in power'. Gillies was especially furious that a member of the Baltic *Free Corps* had been permitted to broadcast to Germany since it was this group who had murdered Rathenau. Campbell Stuart then 'pointed out to him that even political assassins could conceivably be useful for our purpose and calmer waters were reached when Mr. Gillies was reminded of the excellence of Philip Noel-Baker's broadcasts'.

Vogel then gave the SPD's view on British propaganda to Germany. They were recorded in detail. He argued that BBC propaganda attracted far more attention than French efforts because of the truthfulness of the BBC's news bulletins.[23] He also said that the names of British Labour party leaders were known to German trade unionists and Ollenhauer added that German listeners would be interested in speeches by Attlee and Greenwood. Both Germans were told that Sir Walter Citrine had recently agreed to compile a 'trade union leaflet' to Germany. Yet when the talk turned to concrete proposals, Campbell Stuart seemed rather reluctant to enter into any commitments. Gillies insisted that the SPD would not co-operate with non-Socialist exiles not even with 'liberals like Demuth or Weber'.[24] Campbell Stuart noted 'the difficulties of getting effective collaboration from émigrés' if this sort of thing were accepted. Gillies's answer to this was, however, that everything should be done through him. 'He explained that he was anxious to set up an SPD committee of his own.' This bureau would then pass on to Campbell Stuart 'its material and its findings . . . and provided it were given sufficient information, it would prepare broadcasts and leaflets for him'.

Campbell Stuart pointed out that there were difficulties about such a proposal since his 'department could not associate itself with any German political group in particular'. On the other hand, he was ready to use the propaganda of, or for, any specific group which in his opinion constituted 'good enemy propaganda'. In other words, he took the not unreasonable position that he would not commit himself to accept everything the SPD produced nor would he allow the SPD to dominate his exile propaganda output.

[23] See above, p. 20.
[24] See below, p. 99.

Vogel and Ollenhauer may not have understood much of this conversation, although it is recorded that they objected to Gillies calling 'his' SPD committee the 'Headquarters of the SPD in England'. Seen as a whole, however, it must have been a most impressive occasion for them. Possibly for the first time in many years of exile they were being given access to some important officials concerned with the war against Hitler and the Nazis. In addition, Campbell Stuart offered them a real chance of influencing affairs. They must have believed that, were they to come to England, many of the frustrations of exile politics could be relieved. German Social Democracy could gain external legitimation from bodies like the British Labour party and the Foreign Office, and internally the SPD leadership in exile could demonstrate to its immediate and its wider membership that it was able to carry on with the work of advancing the aims of the SPD.

Such views were sustained by the events which followed the SPD leaders' return to Paris. Although the international situation grew daily more grave, the relationship with Gillies and the British Labour party appeared to flourish. Not only were the SPD leaders invited to co-operate on the formulation of propaganda policy but they were also encouraged to feel part of the international Socialist opposition to Hitler. The SPD was treated as an equal and it spoke as one. Vogel made this quite plain at a meeting of the Socialist Workers' International in Paris in March 1940.[25] Although the SPD stood side by side with the western democracies in their fight to overthrow Hitler's dictatorship with any means at its disposal, he wished to place on record that the SPD still hoped the German people would of their own accord get rid of Hitler. 'We Socialists have never demanded the liberation of the Germans by means of foreign bayonets.'[26]

Nor did the SPD adopt an uncritical attitude towards the Labour party even if its reliance on it had been considerably increased. In April 1940 Vogel wrote an angry letter to Gillies complaining that cash which had been promised had still not arrived despite its being urgently needed: 'Affairs in the north (i.e. the invasion of Scandinavia) have had a serious effect on

---

[25] SPD *Mappe* 11.
[26] Ibid.

us financially speaking, fewer people are now reading our press and there is also less opportunity for us to work in the Reich'.[27] And in September 1940 Vogel again complained to Gillies about a man whose name was to figure ominously in the years to come, namely Walter Loeb. Vogel was pleased that Gillies had decided to set up a committee to advise the tribunals on internment, but was outraged that Loeb should have been made a member because 'for weeks now, political and other refugees have been disturbed by his statement that it was mainly on his information that the British Government had decided to intern everyone'.[28]

In January 1940 SPD leaders bitterly criticized British propaganda to Germany in a document which once again shows that they felt fully entitled to participate in the formulation of British policy. They claimed that numerous propaganda lines had been underdeveloped—for example, Allied war aims should be formulated with clarity and Germans ought repeatedly to be told that Nazis and not Germans were the enemy. Specific Nazi propaganda claims should be dealt with in detail—the *Lebensraum* theme could be countered by the question 'why has Hitler encouraged so many *Volksdeutsche* to come into the Reich if there is such a shortage of space?'. And when Hitler suggested that as long as the British continued to lock up Irish Republicans, it was in no position to criticize Nazi Germany, he should be reminded of *Gleichschaltung* in Prussia. If Germany demanded colonies, the Germans should be told that Hitler's racial doctrines made them unworthy of them. More propaganda could be made out of Henderson's account of his visit to Karinhall. The BBC should issue immediate denials of Nazi claims and interrupt music sequences with propaganda flashes. Finally, SPD leaders suggested that more use should be made of Noel-Baker's and Greenwood's considerable popularity amongst the German working class.[29]

The significance of this letter does not lie in its suggestions, many of which seem rather feeble, but in the fact that it demonstrates that the SPD seriously looked to the Labour party for aid in influencing British policy towards Germany, and

[27] Ibid.
[28] Ibid.
[29] SPD *Mappe* 44.

further that the British Labour party encouraged and supported the SPD in this endeavour. For example, on 30 March 1940 Gillies cabled Vogel in confidence advising him not to grant Wickham Steed an interview (he was visiting Paris at that time).[30] Similarly on 12 February 1940 Gillies sent Vogel a copy of a memorandum by Rauschning which was being circulated in high quarters in order that Vogel might furnish contrary evidence. Rauschning's remarks were concerned to prove that Hitler had adopted many of the SPD's political ideas and that this had not only made the SPD obsolete but also proved that opposition to Hitler was to be sought not from the left, but rather from the right.

Rauschning's arguments were intriguing and worthy of consideration not simply because the events of July 1944 gave them some credence, but because he was someone taken very seriously both in Britain and America.[31] He suggested that if Hitler were to be defeated a Bolshevik revolution would immediately break out and indeed, even if Hitler were to remain in power, such a revolution could well occur because the radical socialist side of National Socialism would soon appear. There were some Nazi leaders, like Koch in East Prussia or Kaufmann in Hamburg, who believed that National Socialism was a form of radical proletarian revolution. Whilst the mood of the German people was complex it was certainly opposed to war. The military opposition was strong but unable to strike, first because of the Gestapo but secondly because there had not yet been full scale mobilization. The economy was being planned on Soviet lines and Germany was in many senses a Socialist state. As a consequence, Socialist answers to Fascism should be ignored. The *Entente* should promote the interests of conservative forces in Germany and it should not assume that the Junker caste supported the war. Indeed, the military should be encouraged to seize power. Rauschning ended with the ominous warning that a war against Hitler would help only the Soviet Union and that it made sense only to fight the one as long as one fought the other.[32]

Unfortunately the SPD response to this document is not

[30] SPD *Mappe* 11.
[31] See below, p. 151.
[32] SPD *Mappe* 44.

extant, although it is not hard to say what it would be. The central point worth noting is that this threat to the position of the SPD was communicated by Gillies to the party and we know that Gillies was anxious about the conservative opposition to German Socialism. It is thus not surprising to find that on 10 July 1940 Gillies cabled the SPD leaders 'to come to England by any means, possibly through Spain. Visas unnecessary, landing here arranged',[33] or that on 29 October 1940 Gillies wrote to Sander in the same spirit 'I am doing everything possible for our friends in Lisbon. Reasonable maintenance is assured and I have requested that visas for everybody should be telegraphed without delay.'[34] And the FO records show that on 17 December 1940 visas had been telegraphed to Lisbon at Hugh Dalton's express request, 'The Ministry for Economic Warfare are keen to get these men to the UK . . . as they are needed for a special purpose.'[35]

[33] SPD *Mappe* 22.
[34] SPD *Mappe* 68.
[35] FO 371 24419 c 13495. See below, p. 65.

III

The Attitude of the Foreign Office towards German Political Exiles 1939–1941

FOR the German Social Democrats in London to fulfil their mission in the best possible way, it was vital that they should gain some sort of official recognition from the British Foreign Office and that this should be based on British support for the SPD's claims that it not only possessed a respectable history but that it would also have an important future. The SPD sought British acceptance of its intention to lead German democracy after Hitler by stressing its leadership of the German working class. This made the SPD something, it argued, in which the British had a central interest. To have exerted even a modest amount of influence on the formulation of British policy towards Germany would have been hailed as a major triumph for German Social Democracy and a vindication of its exile. Yet, with one or two exceptions, it can be demonstrated that the SPD had in fact practically no influence at all on the development of Government policy.

Is it therefore possible to take the SPD's aims seriously? Were they anything other than the result of wishful thinking? Was it merely ridiculous to suppose that, confronted with the gravest problems of national survival, the British would find time to listen to a handful of German *émigrés*? Should the apparent disinterest in the SPD shown by the Foreign Office be seen as anything other than the logical way of dealing with political has-beens? The answer to these questions is important because through them we can not only come to a better understanding of the problems of political exile but also of the problems of policy-making in wartime.

It can be shown that the SPD's aims can be taken seriously and that it is not foolish to wonder why they did not meet with success. More than this, however, it can be shown that the British Foreign Office did in fact take them seriously until about 1941. Finally, it can also be argued that the ultimate failure of the Foreign Office to pursue this interest in any active way after the end of 1941 may have been a major error.

As we have seen the SPD leadership had been highly impressed by its meetings with British Foreign Office officials in the winter of 1939/40.[1] It will be recalled that these had been arranged by Gillies not only to serve the interests of the SPD but also to serve the interests of the British Labour movement. Gillies hoped that the influence of the SPD would act as a counterweight to the influence of right-wing German exiles on the Foreign Office. These facts and the cordial reception given to the SPD encouraged its leaders to assume that if London became the official party seat, they would have an important part to play in the defeat of Hitler and the creation of a new Germany.

What the SPD could not foresee, however, was that the Foreign Office was to alter its attitude towards Germany and towards German political exiles during 1940 and 1941. As we shall see the factors which brought about this change were complex and often not without irony. The period has been characterized as the 'unwinding of appeasement'.[2] What this in fact entailed, however, is by no means as obvious as it might seem. For 'appeasement' did not simply mean the appeasement of Nazis, it also implied the drawing of a distinction between Nazis and other Germans. Its proponents saw a chance of avoiding all-out war with the German people either by encouraging the anti-Nazi opposition to overthrow Hitler or else by offering attractive terms to those Germans who were thought to be politically neutral. It is therefore not surprising to find that German Social Democrats were treated with interest in the period before appeasement was finally abandoned, and that they were ignored, in public at any rate, thereafter.

Although the policy of Chamberlain and Halifax which, it

---

[1] See above, p. 36.
[2] See P. Ludlow, 'The unwinding of appeasement' in *Das Andere Deutschland*, ed. L. Kettenacker (Stuttgart, 1977), p. 9 ff. See also Sir L. Woodward, *British Foreign Policy in the Second War* (London, 1962) and *British Foreign Policy in the Second World War* Vol. V (London, 1976); Lord Strang, *The Foreign Office* (London, 1955).

must again be stressed, was far more complex than has often been thought, was finally reversed by 1941/2, it did have interesting results, especially during the 'Phoney War' period.[3] Much stress has been laid by historians on the implications it had for British attitudes towards the German opposition to Hitler and 'peace feelers'.

There has been a tendency to suggest that in the period 1939–40 the British government was testing very seriously the idea of a 'soft' peace with Germany, negotiated with 'moderate' elements among the élite groups of the Third Reich. This is an exaggeration, understandable perhaps in view of the glamour attached to 'cloak and dagger' operations.[4] In reality, these contacts were far more circumspect than is sometimes implied and were for more limited and entirely defensible purposes. They were designed to improve Foreign Office knowledge of the German situation and to gain as much influence as possible over political groups which might be helpful to Britain once Hitler was defeated or overthrown. The thinking of the Foreign Office about such matters is of direct relevance to the reception it gave to the exiled SPD and it is for this reason that some of the most striking examples of Foreign Office interest in German exiles should now be examined. It should not be forgotten that the groups that the British encouraged and this includes the SPD, were the opponents of Hitler and the Nazis.

The case of the leader of the Sudeten German Social Democratic party, Wenzel Jaksch, is an informative example.[5] If we examine the way he, a German of whom very few Britons can have heard, was used to counteract the political weight of another exile leader who was internationally celebrated, namely Eduard Beneš, we can see what the Foreign Office might have done for the SPD. Although Beneš was a favoured ally with excellent connections and Jaksch an enemy alien, the latter received far better treatment because he was politically

[3] See I. Colvin, *Vansittart in Office* (London, 1965); Lord Vansittart, *The Mist Procession* (London, 1958).

[4] One glaring example of this is Peter Ludlow's account of the Christie Mission, Ludlow, op. cit., p. 38. This was first and foremost a means of exploring the ways in which German *émigrés* might help the development of British policy. There is no evidence to show that a secret peace was being planned.

[5] Wenzel Jaksch (who died in 1966) achieved considerable prominence in the Bonn Republic. See J. W. Bruegel, *Czechoslovakia before Munich* (Cambridge 1973), pp. 85, 172. Jaksch was elected leader of the party in 1938.

more useful, and because the policy he stood for suited Britain at that time. Beneš wanted Czechoslovakia to be restored in its pre-1938 borders. Jaksch, quite simply, wanted the Sudetenland to be considered a separate entity.

Soon after Beneš arrived in Britain, having fled the Nazis, he asked the Foreign Office whether he might engage in political activity.[6] He was warned 'not to abuse British hospitality'. This unfriendly admonition was followed by a further request by Beneš for guidance on political etiquette. Winston Churchill had invited him to a lunch in his honour and had asked him to make a speech on the rape of Czechoslovakia. Would this also, Beneš wondered, be an abuse of British hospitality? The Foreign Office humourlessly gave its permission and soon after it also agreed to permit Beneš to have a Czech bodyguard because the Gestapo had threatened to assassinate him.[7] Yet later, when Beneš asked to be allowed to set up a Czech legion to fight alongside the British armies, the Foreign Office became highly disturbed.

Gladwyn Jebb, the Private Secretary to the Permanent Under Secretary of State until August 1940, wrote that he had heard Beneš had set up an exile government in Putney in order to try to embarrass or force the British government into agreeing to re-establish Czechoslovakia after the war. 'This is not a good aim and one to be avoided', Jebb stated and another official agreed with him that, 'it would be disastrous and impossible to recreate Czechoslovakia in its former frontiers'.[8] Later on, however, Beneš was permitted both a foreign legion and an exile government which gained official recognition, although only on the firm condition that Czechs were only to 'fight in the West and that no commitment to any political aims' was being granted by this.

Wenzel Jaksch was initially received by the Foreign Office with no great interest. There seemed to be scant information about him and what he stood for, although it was known that the Gestapo had been anxious to get hold of him. It was, however, soon learnt that he was a 'prominent member of the Czech

[6] FO 371.22898 c 9152.   [7] FO 371.22898 c 10944 and c 12055.
[8] FO 371.23058 c 14840.104. See Woodward op. cit. (1976), p. 64. Churchill stated on 30 Sept. 1940 he welcomed the Czech provisional government and that restoration of 'Czech liberties is one of our principal aims'. But full recognition to Beneš was only given on 18 July 1941. FO 371.22898 c 13228.

Social Democratic party whose mission here was to seek aid for Socialist refugees from the area which had been ceded to Germany'.[9] In 1935 Jaksch's party had received 300,000 votes —more than Henlein's Nazis (250,000) and the German Christian Social party (160,000).[10] Jaksch himself was quick to make the best use of the possibilities that were afforded to him in order to publicize his demand for the special treatment of the Sudetenland. He rented an office, ordered headed note-paper, and began to lobby the Foreign Office.

It was not long before the Foreign Office offered encouragement and, more important, began to build him up politically. On the face of it, this seemed odd, not least because his aim of breaking up Czechoslovakia could have been seen as a pro-German act, and Jaksch was disliked intensely by both Czechs and Germans on account of this. Many SPD members found his 'pan-European nationalism' too similar to that put forward by the Nazi renegade Otto Strasser, to whom he was alleged to be close.[11]

Yet when, on 12 January 1940, Jaksch requested that three colleagues from Stockholm be enabled to come to London to assist his work, Roger Makins, the Head of Central Department until the summer of 1942, supported him. He 'hoped very much that these visas would be granted. We are anxious for political reasons to give discreet encouragement to Jaksch. He and his organisation may in due course become of some importance to us.'[12] The visas were granted.Three days later Jaksch made another request. He asked that 175 Socialist Sudeten Germans be allowed to join the British army rather than Beneš's Czech legion, and on 5 February 1940 he demanded that his Sudeten German Office in London be given the same official recognition as Beneš's Czech National Council. Frank Roberts, Makins's second-in-command, was rather doubtful about this. Although he agreed that Jaksch's ideas were 'sound' he did not think that official recognition would do 'either him or the

---

[9]　FO 371.22904 c 5152, 13 April 1939.

[10]　FO 371.24291 c 1823. The Sudeten German Social Democrats were an issue at the Godesberg meeting between Hitler and Chamberlain in 1938. See I. Kirkpatrick, *The Inner Circle* (London, 1959), p. 117.

[11]　As early as March 1938 Fritz Heine had noted that Jaksch liked to 'ape Hitler by making people call him Fuehrer and causing women to get hysterical'. SPD *Mappe* 51; FO 371.24289 c 24291.　　　[12]　FO 371.24289 c 24291.

Allied cause much good'. But despite this the Foreign Office agreed to Jaksch's proposal only to be told by Jaksch that he had now changed his mind and wished his demand to be retracted.[13]

Three days after this event Jaksch again contacted the Foreign Office because the BBC had called Beneš the 'spokesman for the Sudeten Germans'. The Foreign Office acted on this intelligence and asked Beneš to explain precisely what he had meant by this. It received the answer that he could not do so 'because under British law refugees may not discuss this issue'. An official minuted: 'This is of course nonsense. He just does not want to.'[14] On 12 July 1940 Jaksch demanded to be consulted by the British government on every occasion when policy towards Czechoslovakia was being formulated. Makins, whilst accepting that this made matters rather complicated supported this claim: 'We cannot expect allies to be resolute if they are expected to behave like trained seals.'[15]

By July 1940 Beneš appears to have understood the message the Foreign Office intended for him. He offered Jaksch a seat in his council as a sign that he was now prepared to agree that Jaksch had a case. The Foreign Office was pleased. It noted that after all no more than twenty per cent of Czechs in Britain were Czechs in Beneš's sense, the other eighty per cent being Sudeten Germans, Austrian and German refugees, and Jews.[16] As a result of Jaksch's pressure, Beneš even went so far as to offer him six seats. But the Foreign Office still did not believe that Beneš had been sufficiently forthcoming. On 23 August Makins noted that he had been extremely slow in giving Jaksch his due. Perhaps, however,

he will proceed rather more vigorously when he learns, as he will learn, that we do not regard the Sudeten Germans as coming in any way under his jurisdiction until he has made progress with Sudeten German leaders in this country.[17]

There can be no doubt, then, that the Foreign Office was perfectly prepared to build up a German Socialist political exile and by no means for the reasons one might expect, namely that

13 FO 371.24291 c 1823 and c 4607, 26 Mar. 1940.
14 FO 371.24291 c 2030.
15 FO 371.24291 c 6326.
16 FO 371.24189 c 204.
17 FO 371.24251 c 8917.

Jaksch could have helped in the defeat of Hitler and the destruction of the Third Reich, or that he supported a dismembered or pacifist Germany. Indeed, the very opposite was the case. M15 sent the Foreign Office a report of one of Jaksch's speeches which was quite the reverse of submissive. Jaksch had argued that the Sudetenland needed to be autonomous 'to be safe from Czech oppression'. He continued in characteristic tub-thumping manner

the time for demanding rights for the Germans is not yet ripe. But there will be changes in the English government and then it will be possible to make more demands. If England . . . wants a second Versailles so be it. She will learn we will never sign such a peace. Churchill may win the war, but Bevin will win the peace . . . I have many connections with Sudeten and German workers and they all agree on one thing: it is better to have a Hitler than a dismembered Germany.[18]

As Vansittart commented, this speech was an 'excellent illustration of the German problem, for Jaksch, a refugee and a Social Democrat, still puts the unity of Germany above everything else'.[19] Yet the point remains that Jaksch was taken seriously and that to assume that exiles had no place in the minds of Foreign Office officials is clearly quite incorrect. Indeed, these exiles did not even have to mirror British views in their totality to be able to play a role in supporting British policy. Political exile, then, clearly did not mean political death.

It should not be assumed that Jaksch was the only German political exile who was both listened to and allowed to influence British policy. Hermann Rauschning, for example, was encouraged to provide a steady stream of ideas in 1939 and 1940.[20] He had access to Lord Halifax and he was consulted by William Strang on the direction of European politics after Hitler. Even Otto Strasser's opinions were quoted, though Vansittart warned against him ('we can get on quite well without him here') and a colleague suggested that he was likely to be 'a considerable liability at the end of the war'.[21]

[18] FO 371.24388 c 10448.

[19] Ibid.

[20] FO 371.23058 c 20674, 21 Dec. 1939. FO 371.24289 c 225420. Following the publication of his book *Hitler Speaks* (London, 1939) he was held to be a specialist on Hitler.

[21] FO 371.24438 c 13026.

Another German political exile the Foreign Office was concerned with at this time was ex-Reich Chancellor Wirth who now lived in Switzerland. The Foreign Office wished to ascertain whether Wirth believed there was a future for him in German politics after Hitler had left.[22] In 1940 he was sixty-one and there was some suspicion in London that he was keen on making a deal with the Nazis. This matter was considered sufficiently important for a high-ranking British official to be sent to speak to Wirth. On 23 February 1940 a letter was received from Fritz Thyssen the ex-Ruhr industrialist who had helped finance the Nazi party.[23] It discussed peace terms with 'certain elements in Germany which might one day seize power —namely the *Reichswehr* royalist–Goerdeler clique'. Wirth was seen as part of this group by some, and the Foreign Office wished to find out if this was the case and explore the significance of 'peace terms'.

Lord Halifax and R. A. Butler received Thyssen's letter and the Foreign Office suggested using Thyssen himself as a 'line of communication' to this group. Halifax asked Vansittart for his opinion on this. The brief reply came back 'a *Reichswehr* peace would be a disaster'.[24] Yet the Foreign Office decided to press on with its aim of finding out more about the Swiss connection with anti-Hitler groups. In March 1940 the Foreign Secretary approved a plan to dispatch a high-ranking British official to discover whether there was, in fact, any link between political exiles in Switzerland like Strasser or Wirth and resistance movements inside the Reich. This was the celebrated Christie mission, which has been presented in one account as a serious attempt to make peace with Germany.[25] No one in the Foreign Office appears to have seriously believed that even if such exile figures did have links with anti-Nazis inside the Reich, it would be possible to make a peace treaty with them.

Indeed, in a minute on the Christie mission, Vansittart, who seems to have instigated it, wrote:

Now that Christie has left for Switzerland, he would wish me and I would wish to put on record myself, that neither he nor I nor Kn. nor SS

---

[22] FO 371.24786 c 257 (marked 'secret').
[23] FO 371.24386 3 2339. See F. Thyssen, *I paid Hitler* (London, 1939).
[24] For a detailed discussion of peace feelers see *Das Andere Deutschland*, ed. L. Kettenacker, (Stuttgart, 1977), p. 9 ff. See also, *The Diaries of Sir A. Cadogan*, ed. D. Dilks, (London, 1971), pp. 221, 222.   [25] See Ludlow, op. cit., p. 38.

believe there is the faintest possibility of these manoevres leading to any result.[26]

This mission, then, was not a cloak and dagger operation. It was being undertaken to augment Foreign Office knowledge and not to start negotiations on a peace with Germany. As Vansittart concluded in this minute:

We are convinced that neither Wirth nor the generals nor anybody else can or will deliver the goods of revolution. I hope that we shall, for once and for all, abandon all other experiments whether the will-of-the-wisp dances in the name of phantom generals or of a Fat Field Marshall . . . This war must be won and it can only be won by fighting.

Indeed, Christie later reported that he could find no evidence that Wirth was, as he claimed, in contact with General Halder, that Goerdeler possessed real authority or that the attitude of the German people was revolutionary.[27]

This report was shown to Halifax and to the Prime Minister, Chamberlain. The Christie mission, then, was not a surreptitious peace feeler which would cause one to believe that Chamberlain, Halifax, and Vansittart, not to mention others in the Foreign Office, were defeatist and anxious not to destroy the Third Reich. In fact it shows quite the reverse. It proves that the Foreign Office was alert to the political potential of anti-Nazi exiles, it shows that Vansittart believed war was the best means of getting rid of Hitler and that the people the Foreign Office wished to know more about were the opponents of the Nazi regime. It should not be forgotten that Vansittart personally hated Germany and wanted to destroy its national integrity:

80 per cent of the German race are the moral and political scum of the earth. You cannot reform them by signatures and concessions. They have got to be hamstrung and broken up . . . they are a race of bone-headed aggressors . . . We should certainly aim at splitting Germany up if we possibly can.[28]

[26] Kn. was the code name for a British spy known as the Knight. SS was presumably the Secretary of State for Foreign Affairs. Ludlow does nor reproduce or discuss this minute. FO 371.24389 3 34389, 6 Mar. 1940.

[27] FO 371.24389 3 34389, 6 Mar. 1940.

[28] FO 371.24418 c 5304, 11 Mar. 1940. It is bizarre to suppose that Vansittart supported a secret peace. As Lord Noel-Baker recalled he was one of the most consistently militaristic officials in the Foreign Office. Interview in February 1978. See Vansittart's memo circulated in Foreign Office 'the Nature of the Beast' in March 1940. FO 371.24388 c 4229.

The Foreign Office was not only interested in those German political exiles who lived in London and in Europe. There were many leading exile figures in the United States. John Wheeler-Bennett, the personal assistant to the British Ambassador in Washington, interviewed most of them and his reports were keenly read in London.[29] The most prominent of his contacts was ex-Chancellor Brüning whom he 'saw frequently'.[30] Brüning was asked about the effectiveness of BBC broadcasts to Germany and whether there existed an opposition party inside the Reich. There was a 'German Freedom party', he answered, 'but only in the minds of a few émigrés'. He did, nevertheless, claim that the underground organization of the old political parties, his own in particular, 'could be efficient'.[31] In February 1940 he was asked whether he believed it wise for the British and French to consider recognizing a 'shadow government' of German political exiles should an opportunity present itself. Brüning replied that 'the elements within Germany antagonistic to the regime have made it clear that there will be little or no room for any of the refugee leaders of the Weimar Republic with one exception'.[32] The exception was, of course, Brüning himself.

Wheeler-Bennett also decided to compile a dossier on the SPD in exile. It is interesting to note that the Foreign Office seemed infinitely better informed about centre and right-wing political exiles than about those on the political left. It is by no means strange, therefore, that the British Labour party should have been so anxious to compensate for this.[33] But Wheeler-Bennett's report, which the Foreign Office found both 'useful and interesting' was not particularly well-informed.[34]

The SDP (*sic*) underground organisation is small but efficient. It is far more reliable than that of Otto Strasser. Its real leader is not Friedrich Stampfer who is in ill-health and somewhat discredited with the younger and more extreme members of the SDP, but Karl Frank a young ex-officer of the Austrian army who has been in and out of Germany a dozen times since the revolution of 1933 and has really maintained the morale of the organisation.

Wheeler-Bennett went on to add that Stampfer was a figure of 'very great importance although he could not be relied on,

[29] See Wheeler-Bennett, op. cit., p. 67 ff.   [30] Ibid., p. 166.
[31] FO 371.23008 3 8022.   [32] FO 371.24409 3 3063, 28 Feb. 1940.
[33] See above, p. 33.   [34] FO 371.24419 c 3692.

according to Frank'. Frank, who was not, of course, the leader of the SPD, had also warned Wheeler-Bennett against failing to differentiate between Germans and Nazis in British propaganda to Germany and he urged the British authorities too not to use Jewish refugees, or Stampfer, Breitscheid, and Theo Wolff. They should rely solely on the SPD's reports. Hilferding, Hertz, Schwarzschild and Hoeltermann were also all close to the party.

The Foreign Office was very interested in this but one official was surprised that Frank was said to be the leader of the SPD since he was also the 'founder of *Neu Beginnen*, alias Paul Hagen, well-known to us here'.[35] The *Neu Beginnen* group was indeed known about in London. In June 1940 a Foreign Office official, G. M. Warr was told to supervise dealings with it.[36] His contact was Richard Loewenthal who was to show him all their reports and make copies to be sent to Patrick Gordon Walker in Oxford.[37]

One of the questions that was most frequently put to those German political exiles consulted by the Foreign Office concerned the effectiveness of British propaganda to Germany. This was clearly an area where the exiles could be of immense use to Britain. The Foreign Office was closely concerned with the generation of propaganda material, more closely than is sometimes realized. In this it worked closely with the Ministry of Information and the BBC in a relationship that was based more often on the informal interchange of personnel than on institutional structuring.[38] In February 1940 for example Ivone Kirkpatrick the head of Central Department at the Foreign Office, was seconded to the Ministry of Information in order to bring about closer liaison between the Ministry and the BBC overseas service.[39] Thereafter he became Foreign Advisor to the BBC and, in 1941, the Controller of the European Service of the BBC. Similarly, Richard Crossman and Patrick Gordon

[35] FO 371.24419 c 3692.
[36] FO 371.24419 c 7158.
[37] See above, p. 19.
[38] I am indebted to Professor Michael Balfour for showing me the unpublished draft of his MS on British propaganda to Germany. See also, Kirkpatrick, op. cit., pp. 155-9; Wheeler-Bennett, op. cit.; also A. Briggs, *The History of Broadcasting in the United Kingdom*, Vol. 3, *The War of the Words* (London, 1970).
[39] Kirkpatrick, op. cit., pp. 149-52, 156-9. Briggs, op. cit., p. 333.

Walker both became Foreign Office officials who worked for the BBC from 1940 to 1943 and 1944 respectively.

The first broadcasts to Europe had started in January 1938, and the first ones to Germany after the signing of the Munich agreement. These broadcasts were the fruit of a 'gentleman's agreement' between Lord Reith, the Director General of the BBC and Vansittart. The formal relationship between BBC and government was superseded by an informal one in which the BBC not only took orders from the government about what to disseminate but also was furnished with the individuals to produce that material. Sir Campbell Stuart had been charged with this latter task and he set up an official base at Woburn Abbey aided by people of whom 'only very few had experience of Nazi Germany'.[40]

It therefore comes as no surprise to discover that Campbell Stuart wrote as early as 12 October 1939 to the War Cabinet that he was very anxious to utilize German political exiles in London. His subdepartment had been ordered to 'enlist the knowledge and assistance of German refugees because most of them could render some service to the common cause of destroying Nazism'. Some of them were already engaged in active propaganda. He had, he claimed, been unfairly criticized about the sparse results achieved so far by his subdepartment. The reason for this was first and foremost that the refugees in London were individuals rather than organized in a specific body such as a 'shadow government' which would have led to greater uniformity:

opinions which could emerge from the general body of Germans in this

[40] Kirkpatrick, op. cit., p. 148. In 1940 Dalton was given the secret additional title of Minister of Special Operations and Churchill decided to differentiate between overt and covert propaganda to Occupied Europe. Overt was put under the Ministry of Information covert under the Ministry of Economic Warfare (Dalton's Ministry). The activities at Woburn were rechristened 'special operations' and subdivided into two sections. The first called SO1 was headed by Gladwyn Jebb, formerly of the Foreign Office and its task was to organize and equip anti-Nazi resistance groups in Europe. SO2 dealt with propaganda. In July 1941 SO1 became SOE controlled by the Ministry of Economic Warfare and SO2 became known as the Political Warfare Executive called the PWE. It was under the nominal control of three Ministries, The Foreign Office, the Ministry of Information and the Ministry of Economic Warfare. The head of PWE was Bruce Lockhart but Rex Leeper, another Foreign Office official, played a central role in it. For reasons of secrecy PWE was always called PID, Political Intelligence Dept. This has caused much confusion. See Michael Balfour, *Propaganda in War 1939-1945* (London, 1979).

country would obviously be of much greater weight and value than isolated views and so an attempt should be made to secure regular consultation between the principal groups and set up a central organisation.[41]

At the same time, however, Campbell Stuart considered that the deeply rooted love of Germany by 'Aryans and others' and émigré squabbling would make matters difficult at first. And he concluded that he did not 'conceive the fostering of German political causes to be within its sphere'.

On 24 December 1939 Campbell Stuart gave a further progress report to the War Cabinet, which passed it straight on to the Foreign Office. His pride and joy was a 'new venture in the shape of a fortnightly newsletter to Germany called the *Wolkiger Boebachter*'. This title was, of course, a skit on the official organ of the NSDAP, the *Völkischer Beobachter* and it carried with it an implication that Germany was being observed from the skies and the newsletter was indeed delivered by the RAF. He further stated that he had been active in his attempts at 'coordinating the German refugee groups' whilst at the same time 'withholding any encouragement to schemes for the establishment of a German government in the United Kingdom or to political intrigues of a sectional nature'. In addition he had obtained 'a certain amount of intelligence from émigré channels which had already proved valuable in providing matter for broadcasting'.[42]

Thus, although neither the Social Democratic leadership nor William Gillies could fully know it, the auspices for successful co-operation between Foreign Office and SPD in exile seemed excellent. One further reason for looking to the SPD in exile as a helpful assistant in the struggle against Hitler lay in the fact that many of the other German political exiles in London were considered, by the British, to be first and foremost racial exiles, that is, Jews.

In October 1939 Frank Roberts the second secretary (who succeeded Makins in July 1942) warned his colleagues in a memo 'not to become too associated with purely Jewish activities because they are apt to boomerang'. By this he meant that Britain   might make itself vulnerable to propaganda

[41] FO 371.23105 c 16461.
[42] FO 371.23105 c 16461. See S. Delmer, *Black Boomerang* (London, 1962), pp. 15, 16, 18, 35–7, 38, 40, 43, 50.

charges from Germany that 'we and France are fighting for the Zionists'.[43] During this period the Foreign Office was anxious to secure evidence about the Nazis' use of concentration camps which Halifax himself wished turned into a white paper for propaganda purposes. But he did not want only to tap Jewish sources. 'We should not make exclusive use of Jewish sources but show that perfectly good Aryans and German Catholics have also had to suffer.'[44]

Another disadvantage of using Jewish Germans was the curious belief that they spoke German with a Jewish accent which could be recognized on the radio.[45] Lord Beaverbrook as well as Halifax and R. A. Butler were asked to read a report which said that many Germans criticized BBC broadcasts because 'most of the announcers appear to be German Jews . . . recognized by their Jewish accents'. As a result, Germans 'regard the news from London with suspicion and see it as proof that Britain is run by the Jews'.[46] Yet information from Jewish Germans who also happened to be members of political organizations was sought and utilized. The former leader of the Austrian Social Democratic youth movement, Bruno Kreisky, later to become Austrian Chancellor was asked to give his views on the Austrian exiles in London. The Foreign Office was very interested to see that he believed Austrian workers 'in general hated the Nazi system but are not blind to the merits of Nazism'.[47] Richard Lowenthal was another Jewish Socialist to whom close attention was paid. On the other hand, Bernhard Weiss, a prominent Social Democrat in London who had been in charge of the Berlin police force during the Weimar period, was ignored as a matter of policy, perhaps because he had been singled out by the Nazis for especially vicious vilification.[48]

Notwithstanding these problems the Foreign Office tried hard to push Campbell Stuart's work forward, Indeed, it was so anxious to further the use of German political exiles in

[43] FO 371.23105 c 16788.
[44] FO 371.23105 c 16788.
[45] FO 371.24387 c 1424, 27 Jan. 1940. There is no such thing as a German Jewish accent.
[46] FO 371.24387 c 1424.
[47] FO 371.24387 c 2202.
[48] The Nazis nicknamed him 'Isidor'.

London that it was ready to risk offending its major ally the French government. Despite 'French suspicions' the Foreign Office asked Campbell Stuart to proceed with this notion of

an émigré clearing-house which would be a semi-official committee consisting of English patrons of various émigré groups. Gillies would represent the SPD, Sir Horace Rumbold another batch, Vernon Bartlett the Demuth clique, and Mr. Astor the *Neu Reginnen* Group. This committee ought to have a paid secretary and an office and an official now in the Bank of England is being considered for the post of secretary.

The primary aim of this body was to be the 'transmitting of constructive ideas from émigré sources upwards into government departments' and second to supply detailed information about Germany wherever appropriate.[49] It was envisaged, for example, that the RAF would get advice on where to drop bombs. The 'great merit' of the Foreign Office plan was considered to be that the various exile groups would never actually have to meet each other physically, thus avoiding squabbling.

The scheme was never carried out. Roger Makins thought this was because exiles refused to co-operate with each other, but a colleague believed the fault lay with Campbell Stuart. He had ordered the exiles to 'compile a card index and they not unnaturally instantly suspected that they were being put on the compilation of an index to keep them busy'.[50] They believed that Campbell Stuart merely wanted to 'reduce their nuisance value. Ever since the days of Sherlock Holmes compiling a card index has been a recognised way of keeping people quiet.'[51]

The propaganda line that the efforts were meant to confirm was to be found in the view that Nazis and Germans should not be thought to be the same. Hitler and the Nazis should be deemed to have seized power in Germany and to be holding on to it largely through the use of terror. Were the Nazis to be overthrown it was suggested that 'the other Germany' would emerge, indeed it was thought quite feasible that 'other Germans', that is, non-Nazi Germans, might be encouraged by this view of Germany to revolt against Hitler in the near future. Whether or not this line was based on historical truth

---

[49] FO 371.22420 c 16788.
[50] Campbell Stuart was replaced. See below, p. 71.
[51] FO 371.22420 c 16788, 6 Aug. 1940.

or on wishful thinking should not conceal the fact that it was
not only a most sensible propaganda line to take but also that
it gave the German political exiles a very special part to play in
the defeat of Hitler. In the public mind, however, there seems
to have been an ever-increasing suspicion that it was essentially
on an 'appeaser's' view of the German people. It had, after all,
been Chamberlain who on 4 September 1939 had declared
that the British were going to war against the Nazis and not
the German people.

Wheeler-Bennett writes scathingly about the Foreign Office
attempts to uphold this view in the early years of the war.[52]
He not only attacked the historical premises on which the
argument was based but also claimed that it had a damaging
effect on British attitudes towards the war they had to fight.
He claimed that 'various envoys' from political dissidents had
visited London to give 'highly coloured accounts of embryo
conspiracies'. He dismissed the Foreign Office reaction as 'a
primitive, clumsy and amateurish attempt to drive a wedge
between the German people and their Fuehrer'. Many agreed
with Wheeler-Bennett and the rejection of Chamberlain's approach seems also to have been urged by the French.

On 24 October 1939 Sir Eric Phipps the British Ambassador
in Paris wrote to the Foreign Office that the French, 'strongly
disliked' what they saw as the

growing distinction between Germans and Nazis. The British had been
too soft on Germany after the First World War and they should now
openly campaign for the destruction of a united Germany, otherwise
there was a real danger of divergence between the two nations.[53]

Frank Roberts pointed out that it would be a major change in
policy to support the French demand that the British guarantee
to the German people be weakened and it would be 'at variance
with the line we have hitherto taken'. William Strang was apparently more sympathetic to French wishes for he suggested
the French be told that:

we are fighting to preserve our position in the world against the German
challenge, to establish peace and freedom in Europe and to safeguard
western civilisation. It is less dangerous in the immediate future for western Europe that Russia should displace Germany in Eastern Europe than
that Germany should maintain herself.

[52] Wheeler-Bennett, op. cit., p. 152.
[53] FO 371.23058 c 17104.

Strang modified this anti-appeasement stance by adding that 'the real danger to be avoided is a turnover of Germany to Bolshevism'. The British view was that the Nazi regime should be somehow eliminated and that if the German people could be made to do this with a minimum of effort by the British and French, then this should be tried.

Early in 1940 French fears were expressed a second time. They believed that the British policy on Germany was less warlike than it ought to be. They suspected that the British were allowing right-wing political exiles from Germany to play up the danger of a Bolshevist take-over in order to either avoid dealing with Hitler sufficiently firmly or to further their own political fortunes once the Nazis had gone. Roger Cambon, Minister in the French Embassy in London was ordered to go to the Foreign Office to explore British intentions. He had heard that:

there was, in London a German committee to look after the interests of German émigrés but that under cover of their perfectly legitimate activity it was propagating certain ideas. Amongst these was the demand that this committee be seen as the nucleus of a future German government.[54]

The French might, perhaps, have felt less hostile towards British policy if they had known that it was not simply directed towards right-wing figures like Rauschning. Clearly Rauschning and Otto Strasser were far too close to the political traditions of Nazism to offer a safe counter-weight to Hitler. On the other hand a national committee which included the political centre as well as the Social Democrats might be more acceptable.

The Foreign Office decided to ignore French wishes in this matter. Cadogan wrote that he knew 'nothing of the émigrés in this country and their activities and affiliations' and although Halifax said he *did* know about them, he had 'never heard anything about official recognition'. Although the evidence seemed to suggest that there was a political opposition to Hitler which could serve a useful purpose, it was plainly not easy to organize that opposition in such a way that it would be helpful to the Allied cause and, at the same time, inoffensive to those who feared its actual application.

Yet whatever the difficulties, the Foreign Office had still heard nothing to make it abandon its policy towards German

[54] FO 371.23105 c 16961.

political exiles, even if it was forced to devote less time and energy to this than it would have liked.[55] On 27 June 1940 the notion of a 'standing committee' was taken up by a Foreign Office official yet again:

it is part of our present military plans to create, at an appropriate stage, wide-spread revolt in the territories ruled by the enemy . . . German political refugees could play a considerable role in this. Their first aim would be to formulate a policy to counteract the influence of M15 who are tainted with the Nazi propaganda claim that the refugees are fifth columnists and assume all enemy aliens are hostile.[56]

This, he deduced, was nothing more than an admission of failure by M15 since it was meant to investigate individual cases and 'not rely on prejudice'. Furthermore, 'for political considerations more than military ones' these aliens should be enabled to actually fight the German army.

In the summer of 1940 another way of proceeding was pursued by the Foreign Office with the intention of giving some sort of political status to German political refugees in London. It was decided to let the Germans have their own newspaper. The original idea appears to have emanated from a refugee called Mr Lothar who was a director of the publishing house of Secker and Warburg. On 9 September 1940 Lothar asked the Foreign Office to support the setting up of a newspaper in German.[57] Ever since Schwarzschild's *Neues Tagebuch* had been discontinued, he claimed, exiles lacked a focal point. The purpose of such a paper would be to create a 'certain degree of unity' amongst the exiles who needed more encouragement because German nationals living in Britain were 'not treated as allies even though Germany was the first country to be conquered by the Nazis, even before Austria'. What was needed was a 'centrist daily' which should be 'independently financed, neither by the government nor by the refugees'. It would oppose any peace with Germany without the prior destruction

---

[55] Interview with Lord Sherfield (Roger Makins), 8 May 1979.

[56] FO 371.25247 w 8553, 27 June 1940. This fear of fifth columnists had led many exiles to be interned. On 15 July 1940 Gordon Walker wrote to Warr about the foolishness of this and how one *Neu Beginnen* member Waldemar Von Knoeringen had been released only by 'great efforts'. The fact that he was engaged on 'government work' proved that if 'they are to be of use in the future, they must be free to do organised work'. Chatham House 'would support such a plea since they too think highly of the exiles'. FO 371.22419 c 7158.

[57] FO 371.24423 c 13027. See Röder, op. cit. (1969a), p. 132.

of Nazism, and it would serve as a rallying point for the 'elements of a future Germany' at present in London.

The Foreign Office decided to support Lothar. An official wrote that the 'possibilities of collaboration' with exiles was still 'largely being neglected'. Duff Cooper thought the paper might be a success in which case it could be smuggled into Germany. The only two problems he saw were, first, that too much emphasis would be put on the 'socialist opposition' and not enough on the 'Catholic, conservative and bourgeois oppositions' and that there were 'so many Jews in the direction'. Roberts agreed with this point. Yet he supported the idea not least because the fear of invasion had receded and a 'more liberal policy towards the enemy aliens in our midst is now being pursued'.[58] The proposal went forward and the paper was edited by a refugee who later became a celebrated journalist—Sebastian Haffner—although it never achieved Lothar's aim of uniting the exile factions in London. Indeed, the paper was a major cause of internecine strife among the exiles.[59]

In retrospect the decision to support the publication of a newspaper in German for Germans in London was one of the final acts of Foreign Office support for the German political exiles. Whilst it would be true to say that there was no dramatic watershed in the relationship between exiles and Foreign Office and a number of instances of co-operation occurred later on, the 'unwinding of appeasement' accompanied the unwinding of interest in the political use of anti-Nazi Germans. For a variety of reasons it seems to have become increasingly difficult to build support for these Germans into the new scheme of war policy embarked upon with Churchill's assumption of the premiership. Anti-German feeling was one cause. Ever since the fall of France the threat of a Nazi invasion of Britain had become very real. On 16 July 1940 Hitler had ordered *Operation Sea Lion* to commence and although it had not produced immediate British collapse, it had led to the Battle of Britain in the first part of August. On 14 November 1940 the *Luftwaffe* had bombed Coventry. Instances like these, combined with continuing military activity in other fields,

[58] In the middle of June 1940 the great majority of enemy aliens were interned only to be released after the immediate threat of invasion had passed.
[59] See below, p. 86.

produced a new sense of reality. It was inevitable that a people fighting for its life should see the struggle in straightforward nationalistic terms. And in a democracy like Britain the government was forced to pay attention to public opinion.[60]

In addition, the discrediting of Chamberlain and Halifax produced negative reactions to the policies that they had supported. Churchill was a fighter, determined to win the war by fighting and determined to decide the peace by the outcome of that fighting. Even Chamberlain had been forced to alter his views on Germany. In the final days of his premiership he was asked by an MP whether he still believed it was possible to differentiate between Nazis and Germans.[61] He replied that the responsibility for the prolongation of the war was now as much the German people's as their rulers'.

The Foreign Office appreciated this change of nuance. Roberts pointed out that until the end of 1939 the government was always careful to say 'we were fighting the Nazi government. The German people or the true Germany was being misgoverned by the Nazis.' Gerry Young wrote that now the 'German people had been driven by the external pressure of the war into such a measure of internal unity that they will not be appreciably affected by whether we say we are fighting them or their government'. Eden's view was recorded at the very end of the document:

Hitler is not a phenomenon but a symptom, the expression of a great part of the German nation.

By the time the SPD leadership had established itself in London the attitude towards it, then, had changed considerably. In the period between the winters of 1939/40 and 1940/1 the basis on which real co-operation with the exiled Social Democrats could have taken place no longer existed. This shift in British policy was entirely understandable. Whether it was wise is a matter of conjecture.

[60] Polls were showing a hardening of attitudes towards the Germans. Prof. Balfour has some most interesting public opinion poll results for the entire war period, showing the state of public opinion. See below, p. 120.

[61] FO 371.24362 c 6614, 8 May 1940.

PART II
1941–1943

IV

The Establishment of the SPD
in Exile in 1941

HANS Vogel was flown out of Lisbon in an RAF Liberator bomber to an airport near Bristol at the end of December 1940 and Erich Ollenhauer joined him a few days later. The leaders had grounds for optimism about the possibilities that now presented themselves, even if the international situation looked exceedingly grim and the ultimate danger of capture by the Nazis or of death by bombing had by no means vanished. The SPD in exile seemed to have crossed a major hurdle simply by virtue of the survival of two of its most important leaders and it was now being presented with real opportunities for helping to defeat Hitler and influence British policy towards Germany. There seemed every reason to suppose that as long as England remained free it would be possible for the SPD to preserve a distinctive Social Democratic identity and produce policies and proposals for post-Hitler Germany.

The British Labour party, which was by now part of the British Government, offered German Social Democratic security and an introduction to the men who helped form British policy. The British Foreign Office offered them the chance of a measure of political influence on British thinking towards Germany in return for the utilization of exile skills. These two bodies were of course very important ones, for it was through them that constructive political work could take place. It is certainly demonstrable that the prerequisites for such work existed. The only warning clouds on the horizon were the fact that the Labour party wanted to make the SPD its agent rather than its partner in post-war Europe, and the fact that the Foreign Office seemed not very well informed about the SPD compared to other German political exiles.

Internally the first year of exile in London also seemed to produce much hoped-for advances. The SPD leadership began to formulate policies both for the present and for the future and within a few months of their arrival in England, Vogel and Ollenhauer managed to get the leaders of the three non-Communist Socialist splinter groups, the Socialist Workers' party (SAP), the International Socialist Fighters' Association (ISK), and the *Neu Beginnen*, or New Beginning Group (NB) to rejoin the old party in the form of a 'Union of Socialist organisations in Great Britain'.[1] At the same time, the SPD had changed its own attitude towards its political constituency in Germany. It no longer claimed it spoke for an allegedly repressed and hence silent democratic majority of Germans. Instead it robustly stated that the SPD's views should be those the German people *ought* to be holding. Despite the fact that such a shift made the SPD vulnerable in a number of ways, it did give the leadership a wider space in which to manœuvre.

The promise of this first period was never realized. But it should not be supposed that this was due to the fact that the SPD in exile was considered unimportant, or that it was thought to be absurd to try to work together with those who claimed to represent a Socialist opposition to Hitler. There can be little doubt that in this first period the exiled Social Democrats were not only taken seriously as the representatives of anti-Nazis inside Germany but also as the possible leaders of a post-Hitler Germany. To do so was, indeed, sound for a host of reasons which are considered below.

The Labour party was right to offer aid to a fellow Socialist party and it was right to try to utilize its presence in London. To have dismissed the Social Democratic leaders would have been unthinkable because both Labour party (and the Foreign Office) had the duty of formulating policies towards Germany which would be of immense importance. Anyone with the least knowledge of recent German history could not doubt that if democratic conditions were to reappear in Germany, the SPD would have a sizeable number of supporters. The Labour party believed at this time that it was involved in an ideological struggle against Nazism and Fascism rather than in a war between nations and that because of this Socialists

---

[1] See below, p. 91.

and Social Democrats whatever their national origins were to be seen as the comrades of British Labour. It was committed to a 'Socialist Europe after the war'. In 1940 it had stated in a 'Declaration to the German people' that it had 'no wish to destroy' them and that it was their friend.[2] In May 1940, the TUC, in a 'renewed appeal to the German workers' had recognized that its German comrades had been 'violently torn' from the International Trade Union Movement, and it had stated its belief that the Germans had not 'lost their sense of comradeship and loyalty to the principles' of the 'cause of organised workers'.[3]

Now, in 1941, the same approach could be easily discerned by the SPD leadership. In its May Day manifesto for 1941 the Labour party specifically recalled the 'traditions and achievements' of Germany and proclaimed

We do not forget the bonds of friendship which once united us with you. We realise the position in which you were placed when the Nazi dictatorship broke down the defences of your Trade Union and Socialist organisation. We know with what ferocity your Nazi masters have kept the whip hand over you.[4]

And in the same vein the Labour party expressed its criticism of the government's internment policy which had not 'discriminated between anti-Nazis and Nazis' and made no special provision 'for German and Austrian Socialists'.[5] Such ideas were music to the ears of German Social Democrats. They seemed to prove that the SPD was to be the ally of both British Labour and through it the British government in the fight against Hitler and the Nazis.

Vogel's arrival in London was publicly reported. The *Evening News* introduced him, under the caption 'German enemy of Hitlerism here', as 'another powerful enemy of Hitlerism who has been brought to London at the invitation of the Labour party'.[6] The Labour party immediately arranged not ungenerous financial support for Vogel, Ollenhauer, Heine, and Geyer, who it was thought should be enabled to carry out

[2] See T.D. Burridge, *British Labour and Hitler's War* (London, 1976) and Report of Labour Party Annual Conference 1940, pp. 6, 7, 8, 9.

[3] Ibid., p. 11.

[4] Labour Report 1941, p. 15.

[5] Ibid., p. 30.

[6] LPA, International correspondence (Int. Subcmtee) Box 2, 9 Jan. 1941.

their political duties unhampered by money worries. Heine, a bachelor, received £32 per quarter, Vogel and Geyer who were married with no children received £50 whilst Ollenhauer who had a wife and two sons to maintain received £62 per quarter.[7] Vogel and Ollenhauer together rented a house at number 33 Fernside Avenue in Hampstead which was also to become the office of the SPD in exile.

With the arrival of Vogel and Ollenhauer in England, the Executive of the SPD was now established in London. Of course, not all of its former members could join it there. Some had fallen into the hands of the Nazis, others, like Stampfer and Rinner, were in the USA, and there were also those in exile in Scandinavia. But London was the official seat of the Executive, and was recognized as such by German Social Democrats throughout the world. There were about 160 Social Democrats in London. The leading members were a lively group. They included Erwin Schoettle who had been a *Neu Beginnen* leader and went on to be a member of the post-war SPD and an important party leader. Another Social Democrat was Richard Loewenthal, who worked for the *Observer* newspaper before taking up a post as Professor at the Free University of Berlin. He was not the only journalist in the SPD's ranks in London. Heinrich Fraenkel began a long career with the New Statesman and other important journals at this time, and Curt Geyer, who died in England in 1967, also worked in Fleet Street. Bernhard Menne, who died in 1968, returned to Germany, ending up as Editor-in-Chief of *Die Welt am Sonntag* whilst Erich Brost had a most distinguished career culminating in the senior editorial position in the *Westdeutsche Allgemeine Zeitung*. Fritz Heine not only became a leading political advisor to Schumacher and Ollenhauer but was Press Chief of the SPD until the middle sixties and publisher of *Vowärts* until the early seventies.

Willi Eichler had been in London since 1939. During the Weimar Republic he had been Leonard Nelson's secretary and then leader of the *ISK*. In 1946 he became a member of the

---

[7] SPD *Mappe* 140, the Ollenhauers appear to have settled in well, for by Christmas 1941 they bought a cat to symbolize their normality. SPD *Mappe* 18, Vogel and his wife Dina, on the other hand, were always desperately lonely. Neither of them learnt to speak English or to read it well and they had no social life.

First German *Bundestag*, a very prominent figure on the SPD's Executive and the acknowledged author of the *Godesberger Programm*. He died in 1972. Other noted West German parliamentarians and politicians active in London at this time were W. Heidorn, who became an SPD member of *Bundestag* after 1953 and chairman of the *Deutscher Gewerkschaftsbund* of North Rhine-Westfalia and Arthur Levi who became the chief *Bürgermeister* of Göttingen. Hans Gottfurcht went on to Brussels after London, to work for the international Trade Union Federation, and Walter Auerbach became a permanent Under-Secretary of State in the *Land* of Lower Saxony. Julius Braunthal, whose scholarly works on the history of the Socialist International earned him high acclaim, was part of the SPD's circle, though not a member since he was an Austrian; so for a time was Bruno Kreisky who later became Chancellor of Austria. Wenzel Jaksch, the Sudeten German Social Democrat, became a major figure in the post-war movement for German refugee rights.

Other Social Democrats in London during the war were Viktor Schiff a prominent and highly respected journalist on the *Daily Herald*, Walter Lowenheim, 'Miles' of the *Neu Beginnen* group in Prague, Walter Loeb, a lawyer, Carl Herz who had been one of Berlin's mayors in the Weimar era, Fritz Bieligk who had been a journalist for a Leipzig newspaper, Karl Hoeltermann, the former *Reichsbanner* leader, Adele Schreiber-Krieger, Karl Rawitzki, Ernst Froehlich, and O. Kreyssig. Bernhard Weiss, the former Chief of the Berlin police force played little part in SPD activities in exile although he had been a prominent Weimar personality. E. F. Schumacher who worked in Oxford and went on to be Chief Economic Advisor to the National Coal Board and author of *Small is Beautiful* was close to the SPD leadership. Professor Gerhard Leibholz who returned to West Germany as one of the judges of the Constitutional Court and was married to Dietrich Bonhoeffer's sister, was not a member of the SPD but was highly respected by it, as was another prominent jurist Otto Kahn-Freund. Dr Susanne Miller, Eichler's wife who was to become a well-known West German historian, played host to many of these figures on Sundays during the war at their flat in Welwyn Garden City.

Some of these exiles were Jewish in origin and some of these Jewish Social Democrats found themselves in a most difficult political situation. Because of the Nazis' treatment of the Jews they felt great bitterness towards Germany as a whole and sometimes they did not exclude German Social Democratic leaders from a blanket indictment of their home country. Others however, did not regard themselves as Jews but as Social Democrats who had fled Germany for political and not racial reasons. This difficulty manifested itself especially if Jewish Social Democrats attempted to denounce the SPD for nationalism, as in the case of Walter Loeb.[8]

Not only were they provided for in a material sense, but London was also the centre of European exile politics. New openings appeared each day. On 14 February 1941, for example, Attlee invited Vogel to tea.[9] Hampstead Labour party wrote to say how 'glad' the members were that Vogel was in Britain and 'offered to help in any way possible', and the South Hammersmith Labour party sent invitations for tea and friendship.[10]

Ollenhauer's letters offer further evidence of the initial mood of optimism in the SPD. In his first letter from London on 17 January 1941 he wrote to Geyer who then was still in Lisbon that the favourable impression that they had formed in the winter of 1939 was being confirmed. 'Above all, the quiet determination of the British people to fight the war to a successful conclusion has if anything increased.'[11] On 25 January 1941 he wrote to Katz in New York that their decision to come to London had certainly been correct for there was so much that they could do.[12] Apart from internal party matters (dealt with elsewhere) Ollenhauer saw real opportunities being provided by the Labour party and other institutions such as the Foreign Office and the BBC.[13] Heine was told that the possibilities for major political work were far better than in America which was 'another world'.[14] And by the Spring of

---

[8] See below, p. 109.
[9] SPD *Mappe* 68.
[10] SPD *Mappe* 3.
[11] SPD *Mappe* 80.
[12] Ibid.
[13] Ibid.
[14] Ibid., 24 March 1941.

1941 Ollenhauer believed that London was a better centre for exiles to participate in the overthrow of Hitler than Paris had ever been.[15]

Relations with the Labour party were also, according to Ollenhauer, proving successful. 'Gillies is being very helpful', he wrote to Geyer, 'we can work well with William and we see him at least once a week.'[16] The BBC and the British press had made extensive use of material supplied by Heine concerning the internment camp at Gurs. On 1 May 1941 a broadcast to German workers was planned and *Left News* had asked Vogel for an official statement.[17] An invitation duly arrived from the NEC of the Labour party for Vogel and Ollenhauer to attend the 1941 Party conference. The difficulties in publishing the *Sozialistische Mitteilungen* were to be diminished by aid from Transport House.[18]

As early as 20 January 1941 the International Subcommittee of the Labour party had taken official note of the existence of the SPD group in London.[19] At a meeting, on 1 March 1941, attended by Dalton, Laski, Dallas, Walker, and Middleton, as well, of course, as Gillies himself, Vogel and Ollenhauer were invited to hear Attlee speak and on 29 April the Subcommittee officially requested the BBC to allow Vogel to read a May Day message under his own name. And the Labour party was doing everything it could to get Heine, Geyer, and his wife out of Lisbon from where they were still trying to escape.[20]

By May 1941 Ollenhauer believed that the SPD's first 'six months in England have enabled us to do our normal work' aided by the fact that 'we have food, there are no bombs— and the children go to school'.[21]

One of the most hopeful early contacts was the one made with the BBC apparently through the good offices of Richard Crossman and Patrick Gordon Walker in particular. In 1939 and 1940 the Labour party had expressed its dissatisfaction

[15] SPD *Mappe* 60, 17 March 1941.
[16] Ibid.
[17] SPD *Mappe* 45, Victor Gollancz wrote 'My dear Vogel: Forgive me if I say how fine and brave I thought your contribution was.'
[18] SPD *Mappe* 79.
[19] LPA, Int. Subcmttee. mins. and docs., 1940/1.
[20] SPD *Mappe* 79.
[21] SPD *Mappe* 80, 29 May 1941, letter to Heine.

'with the inadequate manner with which the BBC is utilising the services of the prominent Trade Union and Socialist leaders, refugees in this country'. Following representations made in 1941 there had been a 'notable improvement'. Vogel was asked to prepare a document on the 'SPD and the War' which Gordon Walker said was 'of the greatest use to me'. He wanted Vogel to contribute to the BBC's German programme and Vogel was invited to send along a script for a radio talk to German workers. In addition, the BBC was anxious to have Vogel's opinion and that of his colleagues on the thorny question of anonymity about which there was a considerable disagreement in British quarters.[22] Vogel was then invited to submit yet another script for the next week's programme and he was also told that the BBC was planning a weekly evening broadcast from the middle of February to which, it could be inferred, Vogel and the SPD would be asked to contribute.[23] When Vogel had occasion to complain about the reluctance of the BBC to broadcast news about the SPD's exile activities, Gillies immediately contacted Gordon Walker who replied with alacrity, 'If there is ever any problem, do get in touch with me at once. I am always very happy to receive suggestions and criticism from responsible German Socialists and I would very much like to meet you and Mr. Ollenhauer to discuss this.'[24]

There is clear evidence that Gordon Walker and Crossman believed the exiled SPD leaders could not only contribute to the propaganda war which was being waged against the Reich but also that by supporting this effort the British Labour movement itself would gain considerable prestige. On 4 April 1941 Jim Middleton the Secretary of the Labour party wrote to Arthur Jenkins, Attlee's Parliamentary private secretary, that following the 'considerable revision in the matter of German news talks given under the BBC' Gordon Walker and Crossman were now playing a vital role. In addition, Gordon Walker wished Attlee to know that he was 'anxious to compile a panel of satisfactory German speaking people' who were also Socialists, 'of whom there are not many about'.[25] Who could

[22] SPD *Mappe* 4.
[23] Ibid.
[24] Ibid.
[25] Attlee papers Box 16. Transmissions were now made on a Monday and a Thursday lasting one hour on both short and medium wave.

have been better qualified to fill this shortage than Vogel, Ollenhauer, and their colleagues?

In the summer of 1941, however, there was a serious disagreement between the SPD in exile and the BBC. The issue was concerned with the naming of SPD contributors and broadcasters. German Social Democrats were understandably keen for their names to be broadcast by the BBC along with their contributions, first to add weight to what they said and, secondly, in order to let German workers know that they were still alive and working. The BBC flatly refused to agree to this for reasons which do not seem entirely logical. It was argued that if Vogel, Ollenhauer, and others were allowed to use their names, their remarks might be construed as being the official policy of the SPD which was, of course, exactly what the SPD intended. The BBC could not support this because, as Richard Crossman pointed out in a two hour discussion with Hans Gottfurcht, the BBC would not deal with any alien organization but only with individuals, and it would not accept any material that might be construed as being propaganda.[26]

From the SPD's point of view, however, it was precisely because the leadership's remarks were official ones that there was any reason to broadcast them in the first place. They were being asked to produce scripts because of their specialized knowledge as leaders of the SPD. Indeed by refusing to let them use their names the BBC was implicitly accepting their status as SPD leaders. As Willi Eichler appreciated, the distinction between statements on official policy and propaganda was a very fine one and could be drawn somewhat arbitrarily.[27] Vogel and Ollenhauer decided to take the issue up with the Foreign Office and on 8 July 1941 they went to see Ivone Kirkpatrick.[28] He told them that they had to adhere to the BBC's guidelines or not broadcast at all. It is perhaps a measure of their self-confidence that the SPD leaders should choose the latter alternative. On 14 July 1941 they decided they would no longer work together with the BBC on the existing terms despite the BBC's reiterated promise that it would consider each and every script the SPD sent to it.

[26] SPD *Mappe* 23. See also Briggs, op. cit., pp. 333, 427, 431.
[27] SPD *Mappe* 3.
[28] Ibid.

It is hard to know whether the BBC was simply being clumsy towards the SPD or whether deeper motives may have been involved. General de Gaulle was, of course, permitted to make propaganda on the BBC; indeed, as is discussed below, it can be argued that his subsequent political success in France can be in large part attributed to his broadcasts.[29] If the BBC had believed that the SPD's broadcasts were propaganda it meant that it did not believe the party spoke for anti-Nazi Germans. On the other hand if the SPD's broadcasts did not appeal to the German people or a section of them there was little point in letting the SPD broadcast at all. It seems quite possible that changes in Foreign Office thinking towards Germany, explored elsewhere, rebounded onto the SPD's relationship with the BBC. And it should not be forgotten that the relationship between the Foreign Office and the BBC was extremely close where broadcasts abroad were concerned.[30] The BBC's apparent distaste for propaganda, however, did enhance its reputation during the war.[31]

Gordon Walker, however, did have a clear conception of the use to which German Socialists might be put. He seems to have thought that the SPD could supply the BBC with material that might make cracks in German working class support for Hitler, an idea certainly worth consideration. On 15 July 1941 he wrote to Attlee about this after a discussion with Laski.[32] Gordon Walker had compiled a report which demonstrated that the Nazis feared propaganda that was directed towards the German working class. Any reference to the role that British Labour was playing in the government of Britain was especially resented. A fortnight after British broadcasts to German workers had started, Dr Ley the head of the German Labour Front had suddenly taken a personal interest in German broadcasts to Britain in which he began to attack Bevin's alleged lack of Socialism. This did appear to prove that such broadcasts had been most effective and Gordon Walker told Attlee that this was something the Labour party could use to

[29] See below, p. 164. See Briggs, op. cit., p. 499.
[30] See Kirkpatrick, op. cit., pp. 156, 158, 159. I. Kirkpatrick had been appointed Foreign Office advisor to the BBC by the then Minister of Information, Brendan Bracken.
[31] See Briggs, op. cit.
[32] Attlee papers Oxford Box 16.

its own best advantage. 'This might strengthen any demand by Labour for an increased share in the control of propaganda', he suggested. By showing that it might be feasible to try to encourage working class resistance to Nazi policies, the British Labour movement would at the same time be asserting its own political authority.

It seems hard to dispute that the leaders of the SPD in exile could have played a most useful role in such broadcasts. Even if Vogel and Ollenhauer, as individuals, had by 1941 either been forgotten or were now unknown, German workers and others can hardly have failed to respond to the call of the SPD. The mere fact that an SPD still existed and that it was fighting against Hitler together with Britain would have been highly emotive. Crossman and Gordon Walker, at any rate, held this to be true. They both clearly thought that, even if the German working class was not openly hostile towards the Nazi regime, it was certainly unenthusiastic. Both men were prepared to utilize the skills of exiled Socialists in trying to reach German workers over the air.

Confronted with choice between broadcasting anonymously or not broadcasting at all the SPD leadership had initially chosen the latter. But after a little reflection it was decided that the choice had been wrong. On 17 July 1941 Gordon Walker asked the SPD leaders in Crossman's name whether they would write a script on the attitude that German workers should adopt towards foreign forced labour in the Reich. 'We could if necessary give you material on this subject but I imagine that you have enough.' The BBC, he added, wanted either the one or the other to deliver the talk but it would be done anonymously, 'although we would mention the fact that you are German Socialists'.[33] Marius Goring, the actor, was to liaise with them. On 23 July Crossman himself wrote to thank Vogel for the script which he 'liked very much'.[34]

Yet Vogel still refused to deliver the talk so Crossman tried to get Ollenhauer to agree to do so. 'Read by an announcer', Crossman argued, 'it will lack the vigour and personality that are so large a part of such talks.'[35] On 29 August 1941 Vogel

[33] SPD *Mappe* 4.
[34] Ibid.
[35] Ibid.

was asked to write another script for the BBC on a 'concrete subject like the contrast between the wage stop which is enforced and the dividend stop which can be easily got round'.[36] On 14 September 1941 Gordon Walker offered a new possibility:

a number of regular talks certainly at dawn and possibly sometimes in the evening. I would, therefore, be very interested to have your concrete suggestion of subjects. I am hoping to send out fairly soon the one you gave me last Friday. There is just one point, namely the use of the word Nazi. It is our policy to avoid this particular word, instead to use the translation of it like *die Partei, das III Reich, Hitler-deutschland, die Herrscher Deutschlands*, etc.[37]

By September 1942 the correspondence appears to have ceased and there is no record of any further dealings between SPD and BBC. It should however be plain that the SPD leaders in exile did have an important role to play in this area, and that reasonably full use was made of them until the Autumn of 1941, though individual members of the SPD continued to work for the BBC for the duration of the war. Their assets were their knowledge of the German working class, their political authority, and their commitment to the British cause. It is worth noting that by September 1941 they were even being asked to think up their own subject matter. Furthermore, from the British point of view, Vogel and Ollenhauer had the advantage of not being Jewish. As we have seen British propagandists entertained the continual fear that listeners in Europe would recognize and be alienated by Jewish accents.[38]

Another cause of the optimism felt by the German Social Democrats in London was the result of their dealings with the Socialist Workers' International. The SPD had always been most scrupulous in maintaining its links with this body. When Wels had died in 1939 Vogel had been elected to fill his place. The British representative Philip Noel-Baker and Léon Blum for the French Socialists had always exhibited warm comrade-

---

[36] Ibid.
[37] Ibid.
[38] On 24 September 1941 Hugh Gaitskell sent Attlee a report that 'unfortunately a large percentage of refugees available for the BBC is Jewish', Attlee papers Box 16. See also above, p. 56.

ship towards him. Once in London, one of Vogel's first official engagements was with the president of the SWI, Camille Huysmans. Writing to Vogel in March 1941 from the 'Belgian Parliamentary Office, London' he asked him to speak to the SWI on 'the chances for Socialism in the post-war world'.[39] On 8 August 1941, an even more important invitation arrived at number 33 Fernside Avenue in connection with the SWI. Marked 'private and confidential' a note pointed out that the British Labour movement had decided to extend the work of the SWI and that the SPD was to play its part in its political work. Transport House wished to set up 'a consultative committee of all parties accepting the democratic and Socialist principles of the Labour and Socialist International.'[40]

There were problems involved in enacting this decision. But it is important to note that they were solely administrative in nature. At this stage there was not the slightest suggestion that the subsequent distinction between Socialists from enemy nations and Socialists from allied nations would turn the SPD into a pariah. The NEC accepted that there were 'parties in non-occupied territories' with no delegates, exile parties with delegates, and parties in exile which had never had any delegates to the SWI. The NEC had therefore decided:

that in these abnormal circumstances one is obliged to follow a procedure which is also abnormal and in conformity with the initiative taken during the war of 1914–18 by the secretary of the SWI, who is now its President, Huysmans. He, the chairman of the NEC, Green, the chairman of the international subcommittee of the NEC, Dallas and the International Secretary of the Labour party, Gillies will form a preparatory committee of which Gillies will be secretary.

The SPD leadership was very pleased indeed with this news. It had been discussing the role it might play in the SWI since

[39] SPD *Mappe* 11.

[40] SPD *Mappe* 55, 'Because the Labour and Socialist International can no longer function . . . the NEC has taken steps to ensure that the International Subcommittee and International Department shall maintain closest political and social relations with representative members of the continental Labour and Socialist parties now refugees in this country with the object of giving advice which may be invited on special occasions and the coordinating of such activities as may be necessary for the more effective prosecution of the war against Hitlerism and Fascism.' 1941 Annual Conference Report of the Labour party, p. 31; the phrase 'Hitlerism and Fascism' rather than 'Germany and Italy', is significant.

April 1941. Its first intentions had concerned a proposed May Day rally for that year.

By 17 April 1941 Vogel and Ollenhauer had gained the support of a number of internationally accredited Socialist leaders like Levi, de Brouckère, and Schevenels for their plan of making 1 May 1941 a day of 'international solidarity'.[41] Gordon Walker had promised that the BBC would carry Vogel's opening remarks.

A setback to SPD plans occurred on 22 April, however, when Gillies told Vogel that an international May Day celebration would not be possible and if the SPD proceeded with its plans, it might be thought to be 'unfriendly' towards the Labour party. Gillies, it seems, wanted any such initiatives to come from his office and not 33 Fernside Avenue. Implicit in this is the possibility that Gillies did not wish to see the SPD act except as the agent of the Labour party, though no one seems to have taken up this point. Indeed, Labour party, SPD, and BBC did co-operate on this occasion and Vogel's speech was broadcast.[42]

Some members of the British Labour movement believed the SPD leaders had been humiliated. In the Annual Conference of the Labour party in June 1941, Michael Stewart, for Fulham East DLP, (noting that George Dallas, the chairman of the International Subcommittee had urged closer contact with Socialist refugees in London) added 'especially with our German comrades'. He regretted the fact that the 'International Department prevented other socialists from going along' to the SPD's May Day rally. Even Eden had declared that the Germans would 'have a place at the peace table' and so the SPD should be encouraged by British Labour.[43] George Dallas responded by saying that he did 'not exaggerate' when he said that the International Department 'spends infinitely more time and takes greater pains to do everything that is possible to be done on behalf of the German and Austrian Socialists in this country than the Department spends with refugees of other countries . . . there is hardly a day in the week in the

[41] SPD *Mappe* 3.

[42] Ibid. See below, p. 99.

[43] Labour party Annual Conference Report 1941, p. 152. Stewart was to become British Foreign Secretary in the 1960s.

year in which our members inside the party are not meeting
German Socialists'. He flatly denied that Gillies had told Socialist representatives not to attend the SPD rally.[44]

The promise that the formation of an international preparatory committee seemed to hold was, however, never realized.
By 17 June 1942 Vogel had refused to attend any more of its
meetings.[45] The Austrian Socialists in exile followed the same
course of action.[46] Huysmans was a poor President of the SWI.
In March 1942, for example, Vogel was attacked with some
bitterness for having spelt his name with two 'n's' so presumably making it appear German in origin.[47] But the most significant cause of the failure of this committee was the growing
reluctance to accept the existence of a German Socialist opposition to Hitler. This important development is described in
detail elsewhere but its destructive effect on the SWI is worth
mentioning here. Its main result was to ensure that the committee never left its preparatory stage.

The problem was stated well by Julius Braunthal when he
wrote to Huysmans on 30 December 1942 to discuss the full
reconstitution of the SWI.

Two years ago we were all agreed that the time was not ripe. But now
the Russian working class has entered the grand alliance of the united
nations and the tide of the war has turned in our favour. A victorious
end to the hostilities is—however dimly—in sight.[48]

Huysmans' reply on 7 January was that it was a decision to be
made by the British Labour party and 'not by Socialists who
are generally without any mandate and represent only themselves.'[49] Braunthal's view was that Socialist ideals were inter-
nationalist and that the decision to reconstitute the SWI could
be taken if internationally accredited Socialists wished it, not
simply acting on directions of the British Labour party.

Huysmans's reluctance seemed insurmountable. He appeared
to wish to have no dealings whatsoever with any Socialist from

[44] Labour party Annual Conference Report 1941, p. 152, 153.
[45] SPD *Mappe* 140.
[46] SPD *Mappe* 55.
[47] Ibid.
[48] SPD *Mappe* 56.
[49] Ibid.

an enemy nation. When Braunthal pointed out that if this
were the case, Huysmans should know that the original initiative had come from Victor Gollancz, and that he was prepared
to furnish the evidence for this in print, Huysmans replied 'I
do not like your methods of proceeding. You begin with an
action directed against the Labour party in a question wherein
the Labour party has to play an important part. Now stop this
correspondence.'[50]

Five months later Willi Eichler wrote to Schevenels, President of the International Federation of Trade Unions, to ask
for his support in refounding the international. He begged
him to speak out against any discrimination against the German or Austrian Socialist movements: 'That there is a line to
be drawn between allied socialists and enemy alien socialists
is a shocking invention.' Eichler cited the case of the Belgian
Socialist de Man who now supported the Nazi cause and the
case of the 'enormous number of French Socialists who do
the same'. Yet Huysmans, Eichler stated, still believed that
only the 'Germans are bad and not fit to assist in the re-building
of post-war Europe'.[51] But whatever Huysmans had said must
also have been Gillies's view since his presidential brief now
stated that he had to work 'in the closest cooperation and
agreement with the British Labour party'.[52] The SPD's dealings
with the international Socialist community came to nothing.
As we shall see, this was simply another symptom of the
struggles that were being conducted elsewhere.[53]

However, relations between the Fabian Society and the SPD
were consistently cordial, and dealings, unlike those with the
SWI, not only started well, but continued to prove mutually
satisfying. On 31 March 1941 Philip Noel-Baker invited Vogel
to discuss the prospects for informal co-operation with the

[50] Ibid.
[51] Ibid.
[52] Attlee papers, Oxford Box 8. In 1944 Labour party Conference was told
that over the past year, Camille Huysmans had held over twenty-one meetings
'With all Socialists accepting the democratic and Socialist principles of the Labour
and Socialist International'. But no official representative of the SPD was present.
On 8 September 1944 a plan was drawn up for an International Preparatory Committee from which Germans were also excluded.
[53] Report of the Annual Labour party Conference, 11–14 December 1944, pp.
21, 41.

International Bureau of the Fabian Society.[54] He believed
that it was 'in the present situation necessary for us to develop
closer personal contact with the leading representatives of the
continental Socialist parties to consider the future of Socialism in Europe after the war on a wider basis.

In May 1941 Hans Gottfurcht, who ran the 'Trade Union
centre for German workers in Britain' was asked by the
Fabians to speak to them in London about German trade
unionism. On 16 May 1941 Vogel was invited to supply a list
of German Socialists who would be prepared to speak at Fabian meetings throughout the country.[55] Fabians were

constantly being asked by Labour parties, cooperative societies and
trade unions for people who can speak on post-war reconstruction and
conditions in the occupied territories.

Fabian Socialists quite clearly accepted that the SPD in exile
in London not only spoke on behalf of German Socialists but
also that it had legitimate concerns when the reconstruction
of Europe was at issue. It is interesting to note that frank and
full talks with the Fabian Society continued throughout the
war.[56]

Political co-operation between the Labour party and the SPD
in exile did not slacken during the summer months of 1941.
There were bound to be disagreements like those concerning
the May Day celebrations of 1941 and the difference of opinion with the BBC.[57] Such disagreements were, of course,
part and parcel of any form of political activity and had they
not been present it would seem hard to believe that real
achievements could have been gained. Further evidence which
supports the view that important matters were afoot was provided by a proposal made to the SPD by Lord Davies, the
Liberal leader in the House of Lords and a former Parliamentary private secretary to Lloyd George.[58]

[54] SPD *Mappe* 36.
[55] Ibid.
[56] See below, p. 132.
[57] See above, p. 73.
[58] SPD *Mappe* 4.

Davies approached the SPD with the intention of forming
a so-called German National Committee in Britain with the
distinct possibility that this might become a 'German shadow
government' ready to assume power in Germany once Hitler
had been removed. There has been a suggestion that Davies
was inspired to act by the Foreign Office but of this there is
no documentary proof.[59] He offered the use of a country
house in Scotland to assist the deliberations of his proposed
committee. The German Social Democrats were in one sense
extremely interested in Davies's ideas. Here, after all, was a
further area for political work and one which might bring the
SPD considerable public recognition and consequently lead
to some effective exile policy-making. On the other hand,
however, their political independence might suffer by their
joining a body which was not only British-sponsored but was
also intended to include non-Socialists.

Davies said that unlike the position during the First World
War when there had been no German refugee problem, there
were now a number of important exiles in London who could
be utilized as 'potential allies'. He singled out Dr Bernhard
Weiss, the former Berlin police chief and SPD member, as a
case in point. Vogel and Ollenhauer decided not to deal with
Lord Davies directly but to entrust Hans Gottfurcht with the
negotiations. He told them that in the first instance Davies
wanted to form a council of all the German exiles in London,
an idea which appealed to the SPD, only developing this into
a shadow government later on, a prospect which the socialists
found less attractive.

Davies's main aim was to 'utilise the mental abilities and
the man-power of the anti-Nazi German Democrats, both
Socialist and non-Socialist'. They were to play a full role in
the war effort and the struggle for victory over Nazism and
the German war machine. At the same time, however, they
were to apply their minds to the problems of post-war reconstruction, about how co-operation with the victorious powers
might be best brought about, and how a democratic Germany
might be fitted into the future community of nations. Initially
their work should have the status of research rather than
official policy.

[59] For Foreign Office reactions to Lord Davies see below, p. 148.

On 23 May 1941 Gottfurcht told Vogel and Ollenhauer that he believed the SPD should accept Davies's ideas.[60] A national committee of 'free Germans' should not be set up, however, until full agreement about its status had been reached with the British authorities. It should also be pointed out that 'the political exponents of German democracy abroad should not forget that it is the underground fighters inside Germany who have the vital part to play in the destruction of Hitlerism and the construction of the new nation'. This was, of course, an important article of exile faith.[61] Yet having said all this, Gottfurcht urged the SPD to start talks with Lord Davies. Support came from other important quarters as well. Both Gillies and Noel-Baker advised the SPD to negotiate and an influential rank and file SPD member Walter Loeb, who was to play a considerable role a little later on, had been assured by Lord Vansittart himself that the 'British authorities' were anxious to secure the participation of the SPD. Loeb had also been informed by 'one of Mr. Eden's private secretaries that the necessary facilities for this body to work had already been assured'.[62]

Vogel, Ollenhauer, and Schoettle (one of the leaders of the now subsumed *Neu Beginnen* group) disagreed with Gottfurcht, however. They argued that any negotiations with 'bourgeois groups' would be improper before SPD members had agreed on their own policies towards these matters. At a meeting on 13 June 1941 SPD leaders persisted in their opposition. All agreed that what Davies was, in reality, trying to do was 'create a platform for a future German government', and although Vogel was prepared to believe that a 'German representative committee' might prove useful in the longer term, he feared that any 'German provisional government might produce exactly the opposite effect to the one we want'. He appears to have thought that a German government in exile, founded in London under British auspices would simply allow Hitler to claim that those who opposed him were Allied lackeys.

On 24 June 1941 another Executive discussion on the

[60]   FO 371.26559.7961.
[61]   FO 371.26559.c 7108.
[62]   SPD *Mappe* 4. For a further discussion of this point see below, p. 99.

Davies proposals took place in response to a second plea from him. He reiterated his belief that a German national committee should be set up but now urged that Rauschning, the ex-Nazi, should be included as a full member. This was wholly unacceptable to all German Social Democrats and likely to have been equally unacceptable to the Labour party, not least because it had originally hoped the SPD would come to London in order to counteract the conservative influence of men like Rauschning.[63] Vogel, however, did not wish to condemn Davies openly if he could avoid doing so and he asked his colleagues simply to procrastinate. He felt sure that at this stage Eden would not support a national council.[64] And even if he were to do so, so many other difficulties would emerge that the idea was bound to collapse. There were, for example, no real 'bourgeois leaders in exile' and SPD members in Sweden and the USA were certain to object. The only SPD leader to dissent was Willi Eichler who believed they should support Davies 'and damn the consequences'. A council would have one supreme advantage, he argued, namely to end the *Wichtigtuerei*, the self-importance of some émigrés, for it would sort out the sheep from the goats, those who were invited to be members and those who were ignored. 'The British public would thus realise that not every refugee from Nazi Germany should be thought to be a major political figure in his own country.'[65]

On 28 August 1941 Vogel, Ollenhauer, Geyer, and Heine were invited to Transport House for an important official exchange of views.[66] They were there to discuss international affairs with the NEC international subcommittee against the background of Hitler's invasion of the USSR and the Atlantic Charter.[67] It represented yet another real attempt by SPD and Labour party to co-operate and to formulate policies towards Germany and the war with the active assistance of German Social Democrats in exile. The SPD leaders were specifically asked for their views on the imminent likelihood of a revolution against Hitler inside the Reich. The minutes of this meeting provide documentary evidence of the forthright and serious nature of the co-operation that was taking place.

    [63] See above, p. 33.   [64] Vogel was correct. See below, p. 148.
    [65] SPD *Mappe* 4.   [66] LPA, Int. Subcmtee. mins. 1941.   [67] See below, p. 160.

Curt Geyer led for the SPD. He gave an SPD sketch of the general European situation and the situation inside Germany. He pointed out that it had become very difficult to maintain contact with SPD members in Germany and in the area now under Nazi occupation. He did not believe there could possibly be a revolution inside Germany unless it was preceded by a resounding military defeat which would inflict a major loss of prestige on Hitler. Propaganda, however, could be seen as a useful secondary weapon in achieving such a defeat and Geyer personally hoped that militant nationalism now rampant in Britain would not prevent the making of intelligent propaganda. He feared that growing anti-German feeling would make it very hard to offer Germans attractive inducements to lay down their arms or rise against the Nazis.[68]

Harold Laski responded on behalf of the Labour party. They all wanted to know what the SPD could do by itself, what it could do with British aid and whether the SPD believed that a real opposition would be able to form itself within the Reich. Vogel replied that the first difficulty the SPD had to face at that juncture was that of maintaining contact with Socialists in Germany. The leadership was for this reason considering whether one of them ought to go to Switzerland for the duration of the war in order to rebuild their network. The sporadic reports the SPD did receive, however, stated that neutral observers of German affairs had all detected increasing signs of war-weariness. At the same time there was no sign of organized resistance. They should not, Vogel added, ignore the 'historical examples' which were that this would only happen when defeat had been achieved.

Vogel declared that German soldiers had to be made to realize that to continue to fight for Hitler was senseless. Up to the summer of 1941 this had been very hard to put across since Hitler had met with one success after another. But now, in the wake of the Atlantic Charter and the attack on the USSR, things might prove different. Yet Hitler's hold on the German people remained very strong and there was 'no use in nourishing illusions, for propaganda alone cannot overthrow him'. He urged that any threats about the future territorial integrity of post-war Germany should be avoided. Any suggestion that

[68] This was an absolute reversal of his later position. See below, p. 127.

Germany might be divided or dismembered would prove politically counter-productive. This was, of course, a matter which became increasingly important.[69]

Ollenhauer was then asked about his concerns. He agreed that there was, as yet, no indication of any revolutionary situation inside Germany. But he did think that oppositional 'movements' existed. He and his colleagues concluded with a bitter and lengthy attack on the German language newspaper in Britain, *die Zeitung*.[70] The Social Democrats said that it did not represent the views of the opposition to Hitler and that it lacked a proper editorial line or 'colour'. When Labour party leaders suggested the SPD should co-operate more actively in the publication of this paper, Vogel stated he was willing in 'principle' but only if the SPD was permitted to 'exercise its own authority and put forward its own line'.

It is noteworthy that the SPD leaders believed they could be totally honest with the Labour party leaders about their view of the political situation. They made no exaggerated claims about the existence of an opposition to Hitler inside Germany nor did they try to tone down the extent of the Nazi hold on German society. They looked directly to the First World War experience as a guide to the most likely form future developments were to take. Above all they believed that firmness towards the Labour party would not only not be resented but would also be wholly proper. From the SPD's point of view there was no reason why it should not continue to be treated with seriousness as a political ally of Britain in the struggle against Hitler. The questions put by the Labour party, on the other hand, may with hindsight appear rather more ominous. They were all concerned with the state of German opposition to the Nazis. Although the SPD's opinion may have been realistic they were hardly likely to encourage the British Labour movement and as events turned out, this was to prove highly damaging to both Socialist groups.

There was, as yet, no reason to suppose that in its London exile the SPD would not go from strength to strength. All the indications suggested that as long as Britain remained undefeated, the SPD in London could hope to co-operate with

[69] See below, p. 195.
[70] See above, p. 61.

British individuals and institutions and could hope to generate and develop constructive policies for the duration of the war and for the period thereafter. Nevertheless, the SPD was bound to face serious problems and grave threats. The German army was on the offensive and still apparently invincible. Reports of atrocities were beginning to increase. The German air attack against Britain made it easy to despise all Germans and, as the theatre of war widened to include the USSR and possibly the USA, the situation confronting Germany and the SPD in exile could only become more complex. But at the same time the SPD clearly possessed status and a role, and this could only be to its eventual advantage.

The journal *Socialist Commentary* was a special encouragement to the SPD in exile. It was published for the *Socialist Vanguard Group* and was quite widely read by Labour supporters. In its editorials it consistently adopted a generous, though not unrealistic attitude towards them. In January 1941, for example, it declared that if the Germans were to be incited to revolt against Hitler they would not only require precise instructions from Britain but also the promise of a better Socialist future.[71] Before long, a number of SPD members were invited to contribute articles: Minna Specht, Walter Fliess, Wilhelm Heidorn, and Wenzel Jaksch became contributors, but the most prolific was Willi Eichler.[72]

In October 1941 *Socialist Commentary* began a series on 'Germany without the Nazis' urging that anti-German feeling be prevented from clouding policies towards Germany. Throughout 1943 and 1944 it maintained its support of the SPD, warning against Vansittartite ideas which were diverting attention from the main issue 'the breaking of the economic and social roots of Fascism'. At the same time it castigated

[71] *Socialist Commentary* (1934–72).

[72] Over the years he published an interesting and impressive list of articles. He argued for humane and imaginative policies towards Germany but he was not unduly idealistic. In July 1944 for example, he suggested that spontaneous groups of workers would emerge after the war. These groups would spawn future trade unions. It was however important that the International Federation of Trade Unions ensure these groups did nothing to prevent the destruction of Nazism. Eichler's final contribution came in August 1946 by which time he had become editor of the *Rheinische Zeitung* and a member of the new SPD Executive. He argued against the CDU's wish to create a separate Rhineland state.

the Labour party for failing to produce 'a Socialist foreign policy' and failing to work with SPD leaders.

Among themselves, the leaders of the SPD in exile in London were determined to use their time in Britain to rebuild their exile policies and to remodel fundamentally their approach to their future political constituency in post-Nazi Germany. The SPD was now firmly on the side of those fighting Hitler and the German army. However courageously its leaders conducted themselves this raised the serious prospect that an alliance with the enemies of the Third Reich might prove politically damaging later on. The SPD ceased to claim that its views were the actual views of the German working class at the time, insisting rather that its views were what the German working class would themselves hold once the war was over.

Exile policy-making, then, became qualitatively different from the pre-London era. The party leadership no longer believed its duty was to reflect German political attitudes, real or imagined, but to generate new ones which it hoped would be adopted. This was a new form of exile politics and one which appeared to open new doors. At the same time it should not be forgotten that this course did not imply any willingness on the part of the SPD to give up its political independence. It was determined, at all costs, to remain an autonomous body.

The new course of the SPD in exile was made quite explicit in Hans Vogel's very first speech in London delivered in Transport House on 14 February 1941.[73] He pointed out that an 'enormous change' had taken place in the party's situation now that it was in England. Despite the problems and the dangers that their escape from Vichy France and the Gestapo had produced, the leadership was determined to fight back as never before. They were in London because London was the centre of the war against Hitler. Of paramount importance in all this was the SPD's relationship with the British Labour party and with the London-based European Labour movement. Having said this, Vogel went on to highlight the difficulties that now faced them:

our position is not an easy one. We are not represented by any exile government which could be said to be fighting for national liberation from foreign invaders. Rather we are the representatives of the opposi-

[73] SPD *Mappe* 4.

tional minority of the German people whom Hitler has forced to fight with Britain.

Although it was important for the SPD to realize it was now 'an ally of Britain' it must be careful not to abdicate 'its political independence'. This neat formulation had tactical significance for it seemed to make the SPD's activities in London palatable both to British authorities and to its future political supporters. The British were certainly provided with a realistic basis for co-operation, and Vogel stressed that 'anything the SPD leadership can do to end the war or shorten it or to strengthen the opposition inside the Reich' would be undertaken.

The SPD, then, was determined to assist in bringing down the Nazis but it wanted to do this as the representative of an 'oppositional minority' and not as an appendage of allied policy. With foresight Vogel and Ollenhauer did not conceal the strategy behind this formulation. Post-war Germany's readiness to accept Social Democracy would depend on this point, and this was a practical problem which had to be faced. They argued that the SPD should therefore not simply further the destruction of Hitler but that it should devote as much if not more energy to the devising of post-war policies. As Vogel stated:

Cooperation in the fight against Hitler is only one side of our task. The very first thing that we must do, and it is the hardest, is to work for the peacetime conditions that will, one day, present themselves. We have always said that the war was not the best means of getting rid of Hitler. But now that it is here we must always make it plain that whilst we will not accept any of Hitler's conquests, the German people must be given political freedom and the chance of making a new life.

Having carefully laid down the scope for exile politics, Vogel went on to deal with one immediate and pressing matter. He pointed out that the various Socialist splinter groups in London were strongly disapproved of by the Labour party and other British authorities. Labour politicians wished to see a broader socialist coalition emerge under its tutelage. 'Because of this, our gravest disadvantage, namely our tendency to splinter, should be remedied.' With this order Vogel and Ollenhauer embarked on the first major policy decision of the new exile era, the re-creation of a united non-Communist socialist party for Germany.

Whilst this process was going on, the SPD leaders set about making an effective party organization in London based on the party members who lived in the capital.[74] There were about ninety of these, fewer than the number of KPD members of whom there were about three hundred.[75] There were also fewer members in London than those in Stockholm who numbered about two hundred.[76] SPD members were, of course, a much needed source of funds although the amount that was provided was not excessive, £44 in 1941, £60 in 1942, £104 in 1943 and £91 in 1944.[77] But more important than their financial status was their party political function. They provided political and moral support and an audience for the leaders. By and large, the London SPD members were prepared to come to meetings and on important occasions practically the entire membership would come along.

Eighteen meetings were organized in 1942 and they were attended by an average of thirty-four members, and twenty-three meetings in 1943 were attended by an average of thirty-seven members.[78] There was a London Committee of the SPD which met about ten times a year after June 1942 when it was instituted. SPD members who lived outside the metropolis were less interested in organized party activity, only three members wanted to attend a conference which had been especially devised for them on 27 May 1944. There was, it should be noted, some difficulty about defining who was actually entitled to call himself a member, for those who wanted to be members but had not been members of the SPD before 1933 caused embarrassment. They raised issues ranging from infiltration from left and right to whether the exile party had the right to admit new members. In the event it was decided that provided two existing members were found to act as sponsors, new applications would be accepted subject to veto by the Executive.[79]

Membership meetings were taken seriously. They were often held at the premises of the Austrian Labour Club and they followed a traditional pattern. One held on 26 May 1944 provides a typical example. There was a *gemeinsamer Gesang*, an

[74] SPD *Mappe* 4.
[76] SPD *Mappe* 3.
[78] Ibid.

[75] See Röder, op. cit. (1969a), p. 47.
[77] SPD *Mappe* 13.
[79] SPD *Mappe* 4.

official welcome, a piano recital given by a Herta Polemann, followed by a speech given by H. N. Brailsford concluding with a collection for the Beatrice Webb memorial fund.[80] There were also weekend conferences of which one was held in Oxford, and there were special activities for the younger generation which included the setting up of a youth group in 1943. One of Ollenhauer's sons played an active part in this group where he found his future bride. The SPD in exile possessed, then, an existence at grass-roots level which was important not least for maintaining the morale of the leadership.

On 25 February 1941 various German Socialist leaders in London met at Transport House to initiate the negotiations which were expected to lead to the creation of a reconstituted SPD, one which would subsume the various splinter groups that the exile had so far spawned. Vogel, Ollenhauer, and Sander represented the SPD proper, Froehlich the SAP (*Sozialistische Arbeiterpartei*), Willi Eichler the *Internationaler Sozialistischer Kampfbund* (ISK), Schoettle the *Neu Beginnen* group, and Gottfurcht the trade union interest.[81] What emerges as significant from the record of their deliberations is not so much the fact that agreement was reached or that this agreement lasted throughout the war years and remained an important feature of the SPD in the Bonn Republic, but that the actual terms of the debate indicate yet again Vogel's and Ollenhauer's insistence that although the SPD would fall in with the Labour party's wishes wherever it could, it would not do so at the cost of its own political independence.[82]

Vogel declared that every German Socialist would want to work for a common aim which should be seen not as the creation of a new political party but as a *Vetretung*, a representational body to deal with British Socialists and others. At the same time this new body would have to answer a fundamental political question, 'the most basic question of all, namely should we German Socialists play an active role in the actual fighting?' Vogel argued that the answer to this would have to be an unequivocal 'no' not least because the *quid pro*

[80] SPD *Mappe* 13.
[81] SPD *Mappe* 4.
[82] There are few primary references to this in the Labour party archives. See below, p. 142 ff., but a reference was made to this at the 1941 Labour Party Conference.

quo they might really expect, the achievement of a 'special status' with British institutions like the Foreign Office or the BBC could not 'materialise at this stage'. Vogel, then, aimed to create a revitalized SPD, committed to the destruction of Hitlerism whose leaders would do everything to achieve that end in a personal capacity. Publicly, however, they should do nothing which might cost them the support of their future political constituency.

The next to speak was Willi Eichler who utterly opposed going as far as Vogel. He did not believe the time was ripe for the creation of a reunited SPD, or as he put it the creation of a 'single organisational unit'. Nor should they confine their activities to the British political scene—developments in the USA were likely to mean that German socialists should be active there as well. As for war service, Eichler believed that question was academic—'some ISK members are already in uniform', he alleged.[83] Finally he argued that nothing should be done to take decisions which ought to be taken later on, of which he though the creation of a united non-Communist socialist party was one. They ought, in exile, to confine themselves simply to 'practical questions' on which 'cooperation was possible'.

Schoettle argued in a similar vein: there should certainly be 'loyalty and comradeship' in Britain but the fact that in 1940 some Socialists had been interned had created an element of bitterness about Britain which should not be underestimated. A united party which was not 'anchored in Germany' could not be 'created' he added. Froehlich for the SAP, on the other hand, was anxious to support Vogel and Ollenhauer and he believed that 'Socialists could wear the uniform of another country as long as great value is laid on their political independence from that country'. Gottfurcht who represented the trade union interest agreed and suggested the prime task of the leadership of the new party should be to ensure that persons it did not like were excluded from public activity during the war. Ollenhauer summed up for the SPD: it was plain that no agreement could be reached with Eichler's group

[83] Many ISK members showed very great bravery by agreeing to be dropped behind German lines as Allied agents. See J. Persico, *Piercing the Reich* (New York, 1979).

unless it greatly altered its views. In order to see whether this might be possible, Ollenhauer suggested that there should be another meeting in a week. His position was re-stated: a new party should be simply an 'allied force' (*verbündete Kraft*) for 'freedom and culture'—a phrase specifically designed to exclude 'both Strasser and the Communists'. Their political independence would be preserved and they would not take arms against Germany although the party could support the existence of a 'pioneer corps which did propaganda work'.

On 4 March 1941 the various groups met for the second time at Transport House to consider a resolution which Vogel and Ollenhauer had put before them. It read that

because of the special tasks for German Socialists living in England for the duration of the war, the representatives of the SPD, the *Neu Beginnen*, the *ISK* and the SAP declare themselves united in the conviction that the defeat of Hitler and the final victory over German militarism is a precondition for peace and the new organisation of Europe. For this aim we have decided to fight as an allied force on the side of all opponents of Hitler and for the freedom and culture of Europe with all the means at our disposal.

Eichler, however, insisted that the new party's determination not to be seen as a tool of Allied policy be made quite explicit. He also urged his colleagues to make their anti-Communist stance more plain. Both these points were accepted. The final agreed version ran as follows:

The German Socialists in the United Kingdom agree that they are convinced that the military defeat and overthrow of the Hitler system, as well as the final defeat of German militarism and the abolition of the social base of the Hitler dictatorship are vital preconditions for a peace that is permanent, for the reconstruction of Europe and for a Socialist and a democratic Germany. In view of the special tasks which face Socialists in the United Kingdom for the duration of the war the undersigned groups declare their determination, whilst maintaining their political independence, to fight for the defeat of Hitler and his allies and to join with all opponents of the totalitarian forces in so doing.

Two days later, on 6 March 1941, the various groups met again for the third time to sign this agreed version officially and at the same time it was decided that Gottfurcht should be entitled to send a permanent representative to the new body's executive meetings in order to 'signify the close cooperation between free trade unions and the Socialist movement'. The new body, the amalgamation of SPD, *Neu Beginnen*, SAP, and *ISK* into a new leadership structure was to be called 'The Union of Socialist Organisations in the United Kingdom' under the leadership of Hans Vogel. Ollenhauer had wanted Vogel to chair every meeting but had been unable to get agreement on this.[84] The fact that these organizations were in reality now subsumed under the SPD was concealed by the title 'Union'. Although the various groups continued to have both their own executive meetings as well as 'Union' meetings, as far as the leadership of the SPD was now concerned 'Union' business was SPD business and *vice versa*.[85]

This coming together was a very important development for the SPD in exile in London. First, as the English name 'Union' implies, it was designed to facilitate real cooperation between German Social Democrats and the British authorities in order to defeat Hitler. Secondly, for the first time in its history, the SPD was undergoing internal change in order, in part at any rate, to please a non-German Socialist party. This also accounts for the inclusion of a trade union representative. Thirdly, in order to counteract propaganda attacks the political independence of the German Social Democrats was stressed. This independence, it should not be forgotten, did not simply apply to British bodies but to the German Communist party and to the centre and right-wing German exile groups that existed.

The reintegration of the Socialist splinter groups into the SPD caused some ill-feeling amongst German Social Democrats in exile in the United States. This was to lead to a grave crisis between the two in 1945.[86] Many prominent Social Democrats who had sought exile in the USA and who opposed any alteration in the exile structure of the SPD failed to regain the authority in the party that they had possessed before 1941. Interestingly enough, two other opponents of Vogel's and Ollenhauer's dynamism at this time, Hoeltermann, the former *Reichsbanner* leader and Otto Braun in Switzerland suffered

---

[84] The other groups demanded, and got, two chairmen for an important meeting.

[85] To avoid insignificant detail the 'Union' is hereafter referred to as the SPD leadership except where it is important to differentiate between the member groups (which it hardly ever is).

[86] See below, p. 228.

similar political redundancy after 1945. What these leaders feared was that new departures in SPD policy might be made without their being involved in them. Although both Vogel and Ollenhauer strenuously denied that any fundamental changes were being embarked on and that the agreement that had been reached was simply a question of war-time expediency, this was done more in an attempt to smooth ruffled feathers than to depict the reality of the situation.[87]

On 4 April 1941 the leaders of German Social Democracy in exile met once again at Transport House. This time their political role in the war against Hitler was their agenda.[88] They met in an atmosphere of great hope. Their united stance had, they believed, ensured the continued existence of German Social Democracy and the real opportunity of influencing allied policy now lay before them. Ollenhauer outlined their two main tasks, the fight against Hitler and SPD 'participation in the making of a democratic peace' in Europe. The exiles now needed 'recognition in British public life', since this was the precondition of success in their two tasks. This would not be easily accomplished, he stated, not least 'because the British press is largely Conservative. Yet if we produce pieces of work all the time, we will be noticed.' To this end he sought agreement for the publication of a monthly bulletin to be sent to the newspapers and other British institutions and he committed himself to strengthening contacts between the SPD and the British Labour movement.

Since all had agreed that military action against the Third Reich was not something in which the SPD leadership should actively participate, their twin duties lay, first, in the encouragement of internal opposition to the Third Reich and secondly in the propaganda offensive against Hitler. They ought to concentrate, he argued, not simply on providing technical reports but also 'on making Socialist propaganda'. Whatever their connections they must not cease to 'be free to give the Socialist view as to why Hitler's defeat must also be

[87] Ollenhauer wrote to Katz in the USA that the 'keenness to come together' had originated in the splinter groups and not in the SPD proper, that he and Vogel had only agreed to the formation of a joint party leadership because the Labour party insisted on it and that it was simply the 'prerequisite of propaganda work and not a new united Socialist party' (*Einheitspartei*), SPD *Mappe* 80.
[88] SPD *Mappe* 12.

in the interests of the German working class'. Their main problems, he added, were that they were German and thus bound to be suspect and that they 'did not yet possess the tactical advantage of being an allied government in exile'. At the same time the British themselves seemed uncertain about the political shape of a future German state.

Finally Ollenhauer urged his colleagues to support him in agreeing on a basic policy platform. This was that a democratic Germany could exist and that the German working class was 'the best guarantee for its coming to pass'. German workers were to be the foundation of a new German Social Democracy even if they were not its originators. It was, furthermore, vital that Hitler be militarily defeated rather than allowed to sign a 'soft' peace. General agreement was reached on this line.

The next time that the leadership met, on 11 May 1941, the preparation of a basic policy both towards the British authorities during the war and towards the SPD's future political constituency after it was over, was considered in greater detail. Vogel spoke first on this occasion.[89] He had been rather depressed by the reception of his May Day initiative by Gillies.[90] This, he believed, might mean even greater emphasis might have to be laid on post-war plans since co-operation seemed difficult. They might not be allowed to forget that they were 'political refugees whose hosts had defined the extent of their activities for them'.[91] In addition the Nazis had now 'hermetically sealed' the Reich and it was not only impossible to get information out but also impossible to adequately ascertain the views of their 'friends inside Germany'.

For all these reasons Social Democrats in exile needed to have some first-rate plans to show their comrades in the Reich when they returned to offer an account of their war-time activities. 'It is impossible for Socialists inside Germany to discuss the problems of post-war Germany, so we must do it for them.' That work must now be started, he stated, and in 'absolute openness and with absolute freedom of discussion'.

Ollenhauer then made a masterly speech. Even if he had not achieved prominence after 1945 and had not gone on to

[89] SPD *Mappe* 12.
[90] See above, p. 78.
[91] SPD *Mappe* 158.

lead the SPD from 1953 until 1963, it was a speech which assured him of the role of leading exponent of exile politics, if not nominal leader of the SPD. He provided the necessary inner logic for German Social Democracy to maintain itself during exile and he defined the 'dictates of the moment' in such a way as to give political significance to the SPD's deliberations.

There was, he claimed, no real historical precedent for their exile task in 1941 since a 'dictatorial regime' had never been successfully replaced by a 'democratic one'. This must give them considerable freedom of action in exile. They could agree that 'Hitlerism is a vitally important caesura in German political development'. This meant that there could be no return to the politics of Weimar, which had brought Hitler to power nor would the old European nation-state order re-emerge. He doubted whether the USA would remain neutral for much longer and he expected the 'USSR to produce a few surprises too'.[92]

Who would lead Germany after Hitler had been defeated, he asked. They could all agree there was only one possible answer: 'In 1918 the working people were forced to assume power after Germany's defeat. This time it will be the same. This is because they are the only truly democratic force in German politics.' Furthermore, if the SPD assumed power this would preclude the possibility of Germany being dismembered. The argument that the division of Germany would diminish the desire for aggressive policies was, Ollenhauer argued, sheer rubbish. Indeed to dismember Germany would simply succeed in producing new nationalist feeling. Ollenhauer concluded with a concrete view of the state of the future German nation:

the new German democracy cannot and ought not to be the democracy of Weimar. Its administration must be thoroughly democratised and its economy subject to strict controls. Splinter parties must be abolished by law and democratic rights should be constitutionally protected so that democracy can no longer be destroyed under the guise of democratic freedom. We German Social Democrats must try to put into practice the experiences of British democracy but in a sensible manner by creating smaller constituencies and abolishing electoral lists.

[92] SPD *Mappe* 179.

This sort of rigorous determination marks a new departure in the development of the SPD in exile. Its robust attitude towards the creation and, perhaps more important, the maintenance of democracy in Germany was combined with an undisguised demand for political power by the exile SPD without compromises either towards adherents of the party inside the Reich or towards exile sensibilities. Ollenhauer concluded by urging the setting up of a number of commissions to deal with specific policy areas like the future constitution, the economic and social structure of the new Germany, its foreign policy, and the intellectual matters the SPD would have to consider in order to create decent cultural values.

Ollenhauer was immediately opposed by two SPD members. The first was Bernhard Menne who later left the SPD altogether to join the *Fight for Freedom* Group because he found the SPD leadership too militant and too self-assertive.[93] The task of exile, he protested, was to ask questions not to provide answers in the form of manifestos. Its ultimate aim must be the 're-thinking of past mistakes'.[94] Furthermore, Ollenhauer had failed to mention the Jewish problem simply because the SPD, for doctrinaire reasons, preferred to think it did not exist. But it did now exist and a failure to understand this led Ollenhauer to be wildly over-optimistic in his view of what might happen after Hitler's defeat. The real fact was that by the end of the war Germany would be a wasteland struggling to remain alive.

Menne was supported, perhaps surprisingly, by Vogel himself. Menne was asking for pragmatism whereas Ollenhauer believed that future political reality would simply fall in with his plans. Although he agreed that Ollenhauer's questions were the questions the British were asking, he implied the British also wanted to answer them without the assistance of Germans. German Socialists were forced to conduct their discussions in a 'vacuum' and although improvements in their status were possible as the May Day celebrations had shown, to believe the British would treat the SPD as a political equal was simply a '*Wunschtraum*', a wishful dream. None of them were likely to be 'Karl Marxs or Jan Masaryks'. In the end, however, Vogel

[93] See below, p. 131.
[94] SPD *Mappe* 179.

withdrew his dissent and fully supported Ollenhauer's proposal.[95]

A few days after this meeting there were two further indications that Vogel's pessimism was unfounded and that the SPD in exile was beginning to achieve a measure of recognition. The first was an overture from two Weimar Liberal (DDP) leaders, Demuth and Weber, both of whom enjoyed high reputations. They wished to join the 'Union'. But because they did not want to become Social Democrats, the SPD leadership turned down their request. They insisted that the exile party should not turn into a 'bourgeois grouping'. The second was the invitation from Lord Davies, mentioned previously.[96]

Discussions about the Davies proposal had barely ceased when Social Democrats found themselves forced to tackle another very difficult political problem. On 22 June 1941 Hitler's forces invaded the Soviet Union and Britain soon moved to offer the Nazis' latest victim full support. The SPD in exile was severely compromised by this dramatic turn of events. For a start, the SPD had based much of its claim to special consideration from British authorities on its historical opposition to Communism. The Hitler–Stalin pact had merely served as further proof of the necessity of offering the German working class an alternative to the KPD and of the blood-brotherhood between Fascist and Communist totalitarians of which the SPD had had first hand experience during the Weimar era.[97]

But the attack on the USSR did not only arouse the suspicion that the British might become less anti-Communist. It also raised the question of the internal relationship between SPD in exile and KPD in exile who had now quite clearly, albeit unwillingly, become allies. The BBC touched on a raw nerve when in July 1941 they asked the SPD leadership to include some pro-Communist sentiments in their propaganda talks to Germany.[98] Ollenhauer felt bound to inform his contacts that it was not true that the 'real opposition inside the Reich is now Communist but is the work of organised labour

[95] Vogel seems to have been very depressed at this time. See below, p. 109.

[96] See above, p. 81.

[97] Nazis and Communists had on a number of occasions united to fight the SPD-led Berlin police forces. See Glees, op cit.

[98] SPD *Mappe* 80, letter to Loeb from Ollenhauer, 25 July 1941.

and the trade union movement, whether Social Democratic or free'.

The Social Democratic leadership met repeatedly after the beginning of operation *Barbarossa*. The first and second meetings on 3 and 14 July ended in disarray. Those who wanted to further a left-wing line, to perpetuate the SAP policy, even caused trouble on the third occasion that the question was discussed, on 17 July. They bitterly resented Vogel's refusal to make 'any positive references to the Russian revolution of 1917'.[99] Vogel answered that unless they agreed to the official SPD line as laid down by himself and Ollenhauer, it would be impossible to produce any policy at all. Eichler offered a draft proposal in an attempt to produce a compromise platform on which discussions could take place and a statement be issued to the press. It suggested the line should simply be that 'the enemy of my enemy is my friend'. The USSR was an ally but 'not because it was a workers' state'. But this was not accepted.

On 15 July Eichler produced a second statement. It was beginning to seem ridiculous that almost a month had elapsed since the attack on the USSR and still the SPD in exile could not decide on its policy. Eichler suggested that they say openly that the USSR was only fighting Hitler because it had been forced into it by Hitler who had 'as usual' broken his word. Yet

Now it is a question of the destruction of Hitler and his system and there are only two camps left in the world, those who support him and those who oppose him. The chances for destroying Hitler and the Junkers, the war industry and the military ruling clique have never been so good.

However, this statement also failed to find support and Vogel finally gained the approval of his colleagues by suggesting that the SPD ignore the USSR altogether and concentrate instead on trying to convince the western allies that the SPD rather than the KPD was their true ally. Thus, on 28 July 1941, when Vogel broadcast anonymously to Germany he made no mention whatsoever of the new and decisive change in the world situation. At the same time he warned his German audience that the 'world is filled with justified disgust at Nazi crimes' and unless the German workers offered some active resistance

[99] SPD *Mappe* 12.

they might be considered guilty by association—'*mitschuldig*'.[100] In this way the SPD leaders hoped to prove to the British they were not pulling any punches.

During the summer months of 1941, however, it became obvious that to ignore the USSR dimension was in fact no solution to the problems it raised. Assuming, as was always assumed, that Hitler would in the end be defeated, the USSR's role in that defeat might, as Ollenhauer wrote to Erich Rinner on 1 August 1941 'lead to the USSR occupying bits of German territory'.[101] Furthermore the KPD had decided to approach the SPD with the aim of creating a popular front type of arrangement. 'The KPD wants to work with us', Rinner was told, 'and this places us in a very difficult tactical position.' The KPD had many members and, more important perhaps, it had an adequate supply of funds. It is hardly surprising to find that the SPD leadership began to realize that it was not only the war that was changing but also the internal political development of German Social Democracy.

On 12 August 1941 Vogel informed his colleagues that the KPD had asked for a meeting. Although they disliked any dealings with the KPD, it was thought wisest to agree to meet. Vogel and Ollenhauer, however, refused to go themselves, and instead sent Menne and Sander as their representatives.[102] They had talks with one of the KPD leaders in London, Heinz Schmidt who, in Vogel's own words, 'offered the SPD a great deal in return for being allowed to cooperate with us'. The SPD were not inclined to permit any such thing for the traditional reasons. First, the SPD believed that what the KPD really wanted was to oust the SPD from the political leadership of the German working class and, secondly, that the KPD was simply acting on Moscow's orders and could not be trusted in any way.[103] It was decided to write to Schmidt that the SPD in exile 'saw no possibility of any cooperation'.[104]

[100] SPD *Mappe* 158. Privately SPD leaders doubted the effect of this sort of harangue on German workers even if the British enjoyed it. Ollenhauer wrote on 1 August 1941 'we believe the internal opposition will not materialise until there are real military reverses. Germany's future depends a lot on whether the Germans do anything to resist but of course resistance in a totalitarian state is very difficult.' Letter to Swedish colleague, SPD *Mappe* 80.   [101] Ibid.
[102] SPD *Mappe* 4.   [103] See below, p. 216 and above, p. 18.
[104] SPD *Mappe* 4. See below, p. 177 ff.

At this stage of the war it was relatively easy for the SPD to cold-shoulder the KPD. The outcome of the war in the East seemed unlikely to become obvious until later, although by December the fact that the USSR was still fighting did not go unnoticed by the SPD. As early as 3 December 1941 Ollenhauer wrote to Rinner that it looked 'as if the Russians will survive this winter and that means Germany has lost the war'.[105] It was becoming quite obvious that sooner or later the SPD would have to decide what to do about the KPD even though at this stage there seemed little sign that it was to become one of the most critical issues Social Democrats had to face. The fact that *Barbarossa* was to prove a turning point in the war against Hitler in more ways than one was only gradually beginning to emerge.

[105] SPD *Mappe* 80.

V

The SPD and the Labour Party:
Good Prospects

UNTIL the autumn of 1941 the prospects for co-operation between the SPD's party and the Labour party in the formation of joint policies and plans for post-Nazi Germany seemed good. But all this proved to be a false dawn of comradeship. By the end of 1942 the relationship between the two institutions had deteriorated to a remarkable degree and, less than twelve months after that, the SPD in exile found itself the victim of a Labour party vendetta which the British Socialists hoped would destroy their erstwhile German friends.

William Gillies who, it will be recalled, had enabled the SPD leadership to come to London, was to become the party's main antagonist. The very person who had most supported the view that the SPD could become the ally of British Labour in the war against the Nazis had changed his own view of it so much that by 1943 he was able to write to two leading Labour members:

There is no doubt whatsoever that there is not much left of the old SPD in Germany. It is for the most part forgotten. And as regards the SPD leadership over here—it had better be left alone than brought artificially into action. I am constantly on the alert against any such foolish policy. The HQ regards these émigrés as individuals and does not accept that they are representatives of a party. The Germans' spirit is not really democratic. They are too easily led, much too prone to follow any warlord to the conquest of their neighbours' lands. An insignificant part of the SPD leadership became exiles ... and there is not much basis for the opinion that they will have any influence in Germany after the war. Their principle function now is political controversy. They are not a party. The SPD has not existed for ten years.[1]

[1] Labour Party Archives (LPA), International Department, Middleton papers (M), Box 9. Letter to S. W. Smith, Gen. Sec. of Nat. Fed. of Prof. Workers and R. J. Davies MP, 21 Sept. 1943 and 6 May 1943.

The arguments which had brought the exiled SPD to London and which had offered it the means of continued survival as far as external legitimation was concerned were now being negated.

Those halcyon days when Social Democrats in exile had been treated as the effective representatives for the 'other', non-Nazi Germany and as the true allies of British Socialism had disappeared. Gillies disputed that the SPD had existed after Hitler's assumption of power, he denied the possibility of representative democracy in Germany after Hitler and worst of all, he no longer saw the exiled leaders as a vital and beneficial link between the pre- and post-war SPD. He now believed the SPD had no present and no future.

His views were to all intents and purposes official Labour policy.[2] Not every leading British Socialist supported him but the vast majority did. What was to become the Labour party's new approach to the war and its new line on Germany and the Germans was first glimpsed in June 1941 at the Annual Conference.[3] James Walker MP, the Chairman of Conference, went out of his way to declare

I am not one of those who in this crisis can separate the German people from the German government. The German people are just as responsible for the acts of their government as the government is itself.

This was certainly not an expression of crude jingoism but in the language of the Labour party, Walker's remarks represented a very significant change of emphasis. One immediate result of this was that those in the Labour party who wished the war to be seen as a struggle for a 'Socialist Europe' rather than the destruction of Germany were now banded together with pacificists and those who for one reason or another had never wanted to fight Hitler. Lucy Middleton, the wife of the Secretary of the Labour party, understood this and denounced Vansittartism by name arguing that it would produce a 'Tory peace'. Another delegate opposed the concept of 'fight to the finish' and asked 'have we nothing better to offer than the idea that all Germans are rascals?'

Such views, however, were dismissed as the views of 'Peace

[2] See T. D. Burridge, op. cit.

[3] Annual Report of the Labour party Conference for 1941, pp. 109, 110, 131, 141, 144.

Aimers' in Walker's words, which would lead to a 'negotiated peace' with the Nazis. The attempt to get the Labour party to commit itself afresh to a 'Socialist Europe after the war' by spelling out Socialist war aims was thrown out decisively by Conference. Two and a half million members voted for a hard line and only nineteen thousand voted against it. Thus the changes which the SPD had to cope with as far as the British Foreign Office was concerned were mirrored in its dealings with the Labour party. The causes of these changes have little in common, however, other than the readiness of both institutions to bow to public opinion which was, after 1940, becoming increasingly Germanophobe.

In the Labour party's case these changes were not only more complex but also less easy to justify. The Foreign Office could always claim that its rejection of the German political exiles was in the best interests of Britain at the time and that whatever long-term sense there might be in furthering the SPD, the short-term constraints made this impossible. The Labour party, on the other hand, ought to have found such Machiavellianism less comfortable. The concept of international solidarity and comradeship was part and parcel of the Socialist vision of a better international order. It must have had a special interest in promoting Social Democracy in Europe, indeed it ought to have insisted the Foreign Office help it to do so. A Conservative-led coalition government might not want to see the left prosper in the post-war world, but for British Labour this should have been a primary war aim. Finally, the Labour party's understanding of the nature of working class politics should have made it clear that there would always be a demand for a working class political movement in Germany after Hitler.

What led the Labour party to alter its attitude towards the SPD in exile? A recent study has seen this question as part of the wider problem of Germany, a 'pivotal issue' for British Labour.[4] This was, it is claimed, largely conditioned by 'ideological factors'. As the war continued, so the argument runs, the British increased their hatred for the German people and this hatred was transferred on to all things German and so, by implication, on to the SPD in exile. The only people to

[4] Burridge, op. cit., pp. 13, 23, 24, 84, 105, 106, 118, 120, 164, 167.

oppose this movement it is suggested were those on the left wing of the party whose own ideology was based very firmly on the Labour policy pursued until 1942/3, that there was a fundamental distinction between the German people and their Nazi rulers. It is, however, often difficult to differentiate between ideology and pragmatism in politics. One man's ideology can be another man's pragmatism. Was it ideological for Labour leaders like Laski or Noel-Baker to argue that socialism would re-emerge in Germany and that the Labour party should assist it? Was it pragmatic to believe, as Gillies and Dalton came to believe, that Germans would never produce representative democracy again?

The examination of the relationship between Labour party and SPD after the summer of 1941 sheds light on important questions like these. Whether that relationship was conditioned by Labour's view of Germany, or whether Labour's view of Germany was conditioned by its experiences of the SPD in exile is a complex matter but it is certain that the two factors were an integral part of the same political attitude. The evidence suggests that the closer the Labour leaders probed SPD policy the less kindly they took to the resurrection of German Socialism. Indeed, it is more revealing to examine the instances of conflict between the two groups as an example of Labour policy than to dwell solely on so-called ideological or left-wing doctrines when describing the change that came over British Labour.[5]

One major cause of Labour's disenchantment with the SPD in exile was implicit in Gillies's conception of its role expressed as early as 1939. It will be recalled that Gillies in his conversations with Gladwyn Jebb had argued that although the Labour party would want to see German Social Democrats in power after the end of the war against Hitler, they were to be the expression of British interests and not the expression of what the German people might or might not want.[6] The SPD was to be the agent of British Labour. Although the consequences

---

[5] This is perhaps why Burridge's work is less satisfying than it ought to be. Germany was not simply an intellectual problem for British Labour, it was also a practical one which defined Labour's internal concerns as well as its attitude towards the Conservative party.

[6] See above, p. 36.

of this were at that time not clear, they led to a number of difficulties. For, as we shall see, it was not possible for the SPD to go beyond a certain as yet undetermined point in agreeing to be subservient to British interests.

Two weighty and specific issues became the testing ground for the SPD's intentions. The first was the visit to England in the autumn of 1941 by Friedrich Stampfer and the second was the internal war against the SPD leadership led by Walter Loeb and Curt Geyer. The initial rumblings of this latter affair could be heard as early as the summer of 1941 but it was only towards the end of the year that the Labour party appeared to comprehend what was afoot. In both cases the future of the SPD was the focal point of the matter and in both cases the argument was not, as has been alleged, concerned with émigré quarrelling but with the survival of German Social Democracy, its external legitimation and its internal preservation.[7] These two cases, it will be suggested, were to prove to the Labour leadership that the party's view of the SPD needed to be altered. Whether the Labour leadership simply found what it wanted to find is difficult to say but the indications are that these cases were a decisive factor in the formulation of Labour policy.[8]

Walter Loeb was a Jewish Social Democratic exile in London who had been politically active in the earlier part of the Weimar era when he had been president of the state bank of Thuringia from 1922–4 and because of this post, a deputy member of the *Reichsrat*. He appears to have been a gifted and intelligent man with considerable funds at his disposal but had not otherwise been very prominent in the SPD on a national level. He left Germany for Holland and then came to England. What raised his status from rank-and-file exile to a political figure in his own right was, quite simply, the support

---

[7] Röder, op. cit., (1969*a*), underestimates the seriousness of these incidents, p. 139 ff. He makes no mention of the interviews with Labour leaders.

[8] The Rt. Hon. the Lord Noel-Baker recalled in a letter of 8 Nov. 1977 that it was not the pro-SPD Labour party leaders who went against official Labour policy 'but Dalton, Willy Gillies and the gang of Vansittartites who were in conflict' with it. Burridge, however, does show quite clearly that it was during the Labour party annual conference of 1943 that official policy became anti-German and that until that time 'the distinction between Nazis and Germans remained a cardinal tenet of all official Labour party pronouncements', op. cit., p. 24.

he received in Britain for his views of the SPD's future. This, as he argued in a letter to Hans Vogel on 8 November 1941 (a day with associations for the German Labour movement), was that the future SPD 'must wholly subordinate itself to the military power of the victorious Allies and must be founded solely on it'.[9] He had been advocating such views since 1939 and attracting official SPD anger because of them. In December 1940, for example, Viktor Schiff wrote to Lord Vansittart that Loeb's views were not shared by any other SPD member. There was only one person in German oppositional circles who supported Loeb's request that Vansittart be at the negotiating table when Germany sued for peace, and that was Loeb himself.[10] Loeb's desire to reform the character of German politics was conditioned by the fact that he never wished to return to Germany.

However despite the lack of SPD support for Loeb's views, Vansittart continued to encourage him. In June 1941 he wrote to his colleagues in the Foreign Office that Loeb was a man to be listened to. 'In my view Loeb is the most intelligent, reliable and resolute of the Germans in this country.'[11] And indeed Loeb was listened to increasingly after the SPD arrived in London and not only by those who already supported his views. A number of well-known Social Democrats in London, of whom the most prominent was the Executive member Curt Geyer, began openly to sympathize with his attitude. Loeb and his friends were able to put the SPD leadership in a most awkward position. The more they argued that the SPD had been too nationalistic and should now simply carry out Allied policies, the more the Social Democratic leadership demanded political independence and the more Loeb's thesis seemed to be proved. He and his followers were convinced that the SPD had always pandered to German nationalism. If the SPD leaders agreed, they would have to cease independent political activity. If they disagreed they laid themselves open to abuse. Loeb, Geyer, Menne, Bieligk, Herz, and their associates were, therefore, able to open a number of wounds. They began by

---

[9] SPD *Mappe* 42. Noel-Baker believes Loeb was a Nazi agent which is perhaps not as preposterous a suggestion as it might sound, interview 27 Nov. 1977.

[10] FO 371.26559 c 6873.

[11] FO 371.26559 c 6873.

questioning the SPD's national allegiance and ended by criti-
cizing its role during the First World War and wishing to abro-
gate its right to a political future.[12]

As early as May 1941 Vogel had been confronted with
Loeb's disruptive arguments. Loeb complained to him about
fellow SPD members attacking him and stated this was because
they were anti-Semitic.[13] Vogel confided in the SPD's feminist
leader Herta Gotthelf that Loeb's incessant squabbling was
bringing the party into disrepute.[14] And in September 1941
Vogel complained to an American comrade that Loeb forced
him to 'turn into a Kindergarten teacher if not a regimental
sergeant major'.[15] The important consideration, in Vogel's
view, was that if Social Democrats spent their days quarrelling
like Loeb, they would not be accepted later by those Socialists
who had stayed in Germany. The SPD leadership had the responsibility, Vogel said, of ensuring the exile did not become
a *Sauhaufen*. Pigs like Loeb made this difficult.[16] On 3 November 1941 Loeb delivered a controversial attack on the prominent Social Democrat Friedrich Stampfer who was visiting
London at the time. Stampfer had declared that the 'majority
of the leading proponents of German culture' were in exile.
Loeb claimed that in fact there were only two or three non-
Jewish scientists and four non-Jewish artists in exile.[17]

Loeb's attack on the authorized version of the SPD's exile
role was renewed a few days later on 8 November 1941. Hans
Vogel had suggested that the SPD in London should produce
its own membership cards. The purpose of this was to ensure
the independence of Social Democrats from other organizations and to stress their freedom to make their own policies.
Loeb argued that this was quite unacceptable.[18] Britain was
at war with Germany and so the SPD, he said, should simply
rubber-stamp Allied policies on Germany. To do otherwise
was at best to 'play at party politics', and at worst to commit
the SPD's old sin of 'party-patriotism'. By the latter Loeb

[12] See Röder, op. cit. (1969*a*), p. 146.
[13] SPD *Mappe* 73.
[14] SPD *Mappe* 120.
[15] SPD *Mappe* 120, 19 Sept. 1941.
[16] SPD *Mappe* 120.
[17] T. House M. Box 9.
[18] T. House M. Box 9.

meant that SPD was succumbing to blind party allegiance which had in the past forced it to commit numerous mistakes by putting party advantage before anything else. Loeb declared that Ollenhauer was staking his entire future career on his ability to take the SPD through the war without any self-analysis so that he and the SPD could lead Germany after Hitler.

Loeb went on to suggest that the first thing the SPD should do was examine its past mistakes. It should admit that its policies before 1933 had been entirely wrong. To be accused of Vansittartism, he continued, was no accusation at all. Vansittart's notions were in the long term likely to prove beneficial to Germany. Both he and Loeb believed that post-war German politics would have to be constructed by non-German bodies and that it was only by accepting this that the SPD could have any political future at all.

It is a measure of the seriousness with which the SPD viewed Loeb's actions that Vogel tried to reach an understanding with Loeb.[19] He pointed out that the SPD in exile had a mandate to represent the party. After the war they would be accountable to all their comrades for the way in which they had fulfilled this task. Any deep and bitter rift would ruin their standing and make a return to German political life impossible. Loeb had made it plain that he did not wish to return to Germany after the war but the rest of them were intent on going back.[20]

Loeb's critique of the party was more than mere exile bickering for it represented a real challenge to the exiled party. He not only threatened its credibility of which unity was such a vital part, but he also made strenuous attempts to destroy the party's established view of itself and undermine its belief in its own traditions. This could lead only to the extinction of the party unless steps were taken to re-assert the authority of the leadership.[21]

Loeb's vision of the SPD's post-war role was given an added impetus by the visit to London of Friedrich Stampfer who had come from New York to help his London colleagues. He

---

[19] SPD *Mappe* 112.
[20] SPD *Mappe* 112, 26 Nov. 1941.
[21] See below, p. 166.

was a very weighty figure in the SPD and a member of the Executive.[22] Apart from a most distinguished career in German politics, he was also a leading exile politician in New York. He enjoyed the most cordial relations with William Green, the President of the American Federation of Labor, indeed it had been Green who had personally asked Roosevelt to grant Stampfer a visa.[23] Stampfer received moral and financial support from the American based Jewish Labor committee. He was the editor of the *Neue Volkszeitung* together with Gerhard Seger and Rudolf Katz.

In his memoirs, Stampfer gives a jovial account of SPD life in exile in London. Continental luncheons were eaten at Schmidt's restaurant in Soho, evenings were spent round Ollenhauer's roaring fire in Fernside Avenue. What Stampfer fails to say about his visit however is that it precipitated a very grave crisis for the SPD. Stampfer claimed he had wanted to achieve two tasks in London. The first was to resume personal contact with the leadership of the party and the second was to help devise a series of radio talks for the SPD and the anti-Nazi opposition.[24] Stampfer had wanted to receive an official invitation from the Labour party but one had not been forthcoming since his visit was an internal SPD affair.[25] Nor was Stampfer coming on behalf of the American Labor movement. As Gillies wrote to Citrine, Stampfer was 'not an interpreter of American Labor and the TUC should not be summoned to meet him'.[26]

Yet Stampfer did have two very important meetings with leaders of the British Labour party, Gillies himself and Dalton. Both these visits had a disastrous effect on the relationship between the SPD and the Labour party and must certainly have been a contributory cause of the breakdown in cooperation between them. To the British, Stampfer's views seemed dangerous and antagonistic. To the SPD, however, his views were a vital expression of the political rights of the

---

[22] Stampfer *Erfahrungen und Erkenntnisse* (Köln, 1957).

[23] Stampfer, op. cit., p. 273.

[24] Ibid., pp. 282, 283.

[25] SPD *Mappe* 44, 21 Aug. 1941.

[26] LPA M. Box 8 Citrine's hand had been pressed into Stampfer's by Green at the 1940 New Orleans congress of the American Federation of Labor. The delegates were delighted but Citrine was enraged, Stampfer, op. cit., p. 277.

exiled party. Whilst it was quite true that Stampfer's gifts lay more in polemical journalism than in diplomatic exchanges, Stampfer and the SPD were not wholly responsible for the friction they generated.

On 21 August 1941 the *New York Times* had published a letter written by Stampfer about post-war policy towards Germany which aroused very wide interest. He argued that the western powers had been only too happy to occupy the Rhineland with troops when a Social Democrat, Ebert, had been in charge of German affairs. When a Nazi controlled the Reich, however, the Rhineland was handed to him on a plate. He urged that such errors should not be repeated a second time and said that if Allied policy was constructed around the precept that the destruction of Germany and the German people was essential, the Germans would simply fight on and on. And as far as a second Versailles was concerned any attempt to disarm Germany unilaterally would be resisted in the strongest terms.[27]

However logical such an argument might seem to be, there was no doubt that it would cause passions to be inflamed. To speak of resistance to Germany's being disarmed was tactless and foolish. Such views might enable Stampfer to prove after the war that he had not been anti-German during it but they would hardly enhance the chances of the SPD helping in the formulation of British policy.[28] Gillies, when first shown this letter in a meeting of the NEC international subcommittee, simply dismissed Stampfer as 'childish' and he told George Dallas that Stampfer's extravagant language was solely motivated by a desire to bring about a political solution to the war. If the war 'were allowed to continue, nothing in England would be left standing'. It was not Stampfer's aim to 'shield Germany'.[29] Yet whatever the underlying reasons for Stampfer's polemics they were in themselves bringing into the open matters which Vogel and Ollenhauer would have preferred to deal with more quietly.[30]

Stampfer went to see Gillies in Transport House on 23

[27] LPA M. Box 8.
[28] Although Stampfer returned to Germany, he never re-entered politics.
[29] LPA Int. Subcmtee mins. and docs. 1941.
[30] See below, p. 167, for SPD leaders' view of unilateral disarmament.

September 1941. His purpose was 'to tighten the connection between the German Labor Delegation in America and the British Labour party'. Stampfer argued that the GLD's claim to political independence was firmly supported by the twenty million Americans of German origin who refused to accept the assertion that 'all Germans are gangsters'.[31] Although American Labor as such did not care in principle which nation won in what it considered simply an old-fashioned struggle, concerned with national boundaries, it did not want 'slave labour to win against free labour' because it could not compete with the former. Thus American Labor did not want a Nazi victory and no less a figure than Green himself wanted the 'SPD to participate in peace talks with Germany' which was the 'ninth point in the Atlantic Charter'.

Gillies then gave his view of the attitude of British Labour. It did not want to have fixed plans about a future peace at this stage. All the British were concerned with was winning the war and avoiding defeat. 'The common people' were sure of only one thing, he added, 'that Germany must not be permitted to do all this again.' Gillies then taxed Stampfer with his *New York Times* letter. Did Stampfer, Gillies asked, now wish to 'agree with Grzesinski, Hamburger and Katz that Germany should not be required to disarm unilaterally?' Stampfer gave an ambiguous response. What was important was that there should be disarmament, he stated, but negotiation was a better means of achieving it than its being unilaterally imposed by force of arms. In addition, Stampfer believed, the German people would themselves want Germany's disarmament. Gillies then asked 'but what guarantee can you give that Germany will not fight again?' Stampfer delivered his reply in terms unlikely to win Gillies's support. There could, he said, be 'no guarantee other than a wise Anglo-American policy'.

Stampfer's views and his means of expressing them were wholly appropriate as the views of a sovereign political party representing its particular national interests. But they were hardly appropriate for the SPD's position in London. This does not mean that the SPD could not hold these beliefs but that it could not afford to express them openly and with as

[31] LPA Int. Subcmtee, 1941.

much acerbity as Stampfer was doing. Gillies quite clearly
believed he detected a most dangerous trend in Stampfer's
remarks which boiled down to the prospect that once Hitler
had been vanquished, the SPD might refuse to do anything to
prevent further extremist takeovers. Gillies asked him why
exiled leaders of the German opposition to the Nazis like
Brüning refused 'openly to attack Hitler'. Stampfer said he
was not responsible for the views of the Centre party. It was
precisely because exiles said such things, Gillies rejoined, that
the BBC's German service was not being run as a refugee
station. In any case the Germans would not take such propaganda seriously until military defeat was certain. Stampfer did
not disagree.

The discussion then turned back towards the SPD. Stampfer
seems to have sensed a desire on the part of British Labour to
bring about a fundamental reform of the German Labour
movement because the history of the Weimar era apparently
proved it had grave flaws. But Stampfer argued, the achievements of Weimar were being ignored. The SPD had accomplished far more than it had been credited with. Gillies disagreed. Weimar had not altered the development of German
politics. After the war he guessed the Germans would blame
Britain for having started the war by going to Poland's defence,
with which Stampfer agreed. Gillies's leading questions were
clearly intended to confirm his thesis. Stampfer, with singular
stupidity, did nothing to prevent this. Nothing was said to
Gillies about the terror the Nazis had used to gain power and
were using to keep it, nothing was said about the SPD's oft
repeated pledge to undertake a fundamental reform of
German society.

Gillies brought the interview to a conclusion by suggesting
in view of all that had been said that it seemed most unlikely
that the SPD would emerge spontaneously after the war was
over. Young Germans were all bound to be Nazis, even if their
parents had been Social Democrats. Nazis would still exist
and private armies would re-emerge to help them gain power.
Everything, in short, would be a re-run of 1918; indeed, social
and political revolutions would take place throughout Europe
including France and Poland. The only sensible policy, therefore, was that Germany's affairs be completely controlled by

a military government of occupation. Surely Stampfer could agree to this. Stampfer, however, could not. 'The only guarantee of peace is an SPD-led government', he declared.

On this note the interview ended and Gillies was left to compile a report for his Labour party colleagues. Needless to say it was not only condemnatory but it also used Stampfer's outspoken opinions as evidence of the unreliability of working with German exiles. They wanted to rehabilitate themselves by proving that they had done something to 'ensure for Germany a peace of reconciliation after defeat. Stampfer obviously expects to find a bitter nationalistic atmosphere in Germany after defeat'. Although Stampfer had been exceedingly careless in his choice of arguments it could be said that Gillies had brought out the worst in him. Gillies's truculence was well-known and it was perhaps unfair to take the things Social Democrats in exile were more or less obliged to say as the final statement of SPD policy.[32] It was hard for the SPD to be a serious contender for post-war political power unless it upheld its justifiable belief in a German opposition to Hitlerism and in a sound future for democracy. But there was little sense in pressing such notions to the logical conclusion because they were irrelevant as far as actual policy-making was concerned. Gillies, his Labour colleagues and Stampfer all refused to recognize this.

Stampfer appears to have realized that his meetings with Gillies had been counterproductive because he decided to side-step the International Secretary and appeal to Attlee directly.[33] On 20 October 1941 he wrote to him that American anti-Nazi organizations and exile bodies were 'now planning a vast and well-directed propaganda offensive against the Third Reich', in very close co-operation with the American authorities and with the active assistance of Colonel Donovan of the OSS, the American Secret Service. He added that in America the existence of exiles from Germany was always given the widest publicity. Unfortunately for Stampfer and the SPD, however, Attlee simply passed this letter to Gillies, taking no apparent interest in the matter.

[32] See below pp. 142, 227. Without exception those who came across Gillies remarked on his unpleasantness.
[33] LPA Int. Subcmtee mins. and docs., 1941.

Yet worse things were awaiting Stampfer, for on 9 November 1941 the *Sunday Times* carried a leading article bitterly opposing his views and urging that Stampfer be deported from England.[34] The article stated that

Britain now had visitors on whose activities we should like further light . . . We are bound to criticise the intimation that Herr Stampfer favours multilateral and negotiated disarmament. He should declare himself publicly and strongly in favour of Article eight of the Atlantic Charter. We request the government to make it clear that there is no room here for any guest who works against our declared national policy. In this light . . . there would be a careful review of the aims and activities of all visitors of enemy origin in a position to influence the expression of British policy.

Article eight of the Atlantic Charter, it will be recalled, dealt with the destruction of the Nazi regime and the bringing of peace to Europe. Although the *Sunday Times* leader did at least accept that Stampfer and the SPD were in a position to 'influence' British policy, it made it extremely difficult for Stampfer to carry on with his political work in London. The article may well have been inspired or actually written by Lord Vansittart who was a regular contributor of features and poetry to the *Sunday Times*. In this case, it must have been intended to destroy Stampfer's credibility as a potential ally of Britain by making English authorities very wary about having anything to do with Stampfer, a point happily taken up and exploited by Radio Berlin.[35] Stampfer's apparent refusal to support unilateral disarmament for Germany was being related to sinister political motives.[36]

Stampfer's final interview with a leading British figure, Hugh Dalton, was not likely to produce positive results as a consequence of all this. The SPD had set great store by a meeting with Dalton. A book he had written for Penguin in

[34] See *Sunday Times*, 9 Nov. 1941.

[35] *Sunday Times*, 23 Nov. 1941.

[36] Stampfer did not reply to the *Sunday Times* on Vogel's advice. Instead he issued a statement which was sent to the press and Labour leaders which was moderate and conciliatory. 'The SPD, he said, 'was ready to collaborate with all forces which intend to redeem liberty, justice and order. The Nazis say that the destruction of Hitler is not possible without the destruction of Germany. As long as Germans believe this they will fight on'. The only antidote to this was SPD propaganda from London, 'Give the exiles the means to speak to the German people. Well directed and mighty offensives by propaganda will hasten the triumph of democracy.' SPD *Mappe* 12, 27 Nov. 1941.

1940 entitled *Hitler's War* had been greatly valued by German Social Democrats. Dalton had praised the 'great achievements' of the Weimar era and pledged his support for a 'bold and generous course' in making a peace with Germany after Hitler.[37] He stated that the Labour party was 'opposed to any attempt from the outside to break up Germany or seek its humiliation and dismemberment' and he specifically said it had been to the 'discredit of Britain and France that they did not go out of their way openly to make friends with the democratic leaders of Weimar Germany and help increase their prestige which was a very grave mistake'.[38] Finally he reiterated his support for the official line by urging that

Any attempt to keep Germany an outcast after this war will fail. The most far-sighted and least dangerous policy is to seek to win the cooperation as our equal partner of a Germany governed by a political system whose aims and needs run parallel with ours.[39]

At their meeting, however, Dalton was no more enthusiastic about Stampfer than Gillies had been and probably for the same reasons. Nothing that Stampfer was able to say to him could dispel suspicions about Stampfer's intentions which had come under even closer scrutiny since the *Sunday Times* editorial. Indeed, privately, even Curt Geyer who was, after all, a very close colleague of Stampfer's, was saying that the real purpose of his visit to London was to explore the prospects for a negotiated peace with the German army and for this a German general had come to London to meet him. Stampfer, he alleged, needed to offer something to the *Wehrmacht* in return and this was the SPD's commitment to the existence of a future German army and its opposition to unilateral German disarmament.[40]

Stampfer tried to get Dalton to agree to co-operate with the SPD in a number of areas of which the most pressing was propaganda to Germany.[41] He noted that the SPD leadership was finding it surprisingly difficult to gain the sort of acceptance it had believed would be forthcoming.

[37] H. Dalton, *Hitler's War* (London, 1940), pp. 22, 148.
[38] Ibid., p. 22.
[39] Ibid., p. 187. This was official Labour policy. See Attlee's speech of 9 Feb. 1940 issued as an NEC document.
[40] Stampfer had a reputation for speaking to German generals. See above, p. 18.   [41] LPA M. Box 9.

He contrasted British reservations with the kind of support the Americans were now offering and he produced transcripts of the radio talks that Max Brauer the ex-SPD mayor of Altona had given on USA radio. SPD leaders in London should be permitted to do the same, on the lines of *Les Français parlent aux Français*. Dalton refused this saying the question of personal talks had been 'much considered but subject to certain exceptions we are against them'. Stampfer pointed out that in America SPD leaders were taken very seriously. Katz, for example, had regular contacts with Colonel Donovan of the OSS. To this Dalton replied that he would raise the matter with his Cabinet colleagues.[42] They would decide whether SPD leaders in Britain should be approached in the same way.

Stampfer then asked Dalton outright whether the British authorities really wanted to work with the SPD in exile, especially in the formulation of propaganda policy. The SPD's willingness to be considered an equal in the fight against Hitler was well-known. Indeed it was precisely because of this that Vogel, Ollenhauer, Geyer, and Heine had been brought to England at Dalton's express wish.[43] Dalton replied that there were now very serious obstacles in the way of such co-operation, of which one was 'Stampfer's own stated opinion on disarmament'. Stampfer said this could not be the real reason. The fact was that the Labour party was allowing itself to be increasingly dominated by Vansittartite arguments which were preventing it from thinking sensibly about policy towards Germany. Loeb for one had been allowed to influence Gillies and the International Socialist leader Huysmans 'out of all proportion to his real political significance'.

Stampfer was fully prepared to accept the first seven points of the Atlantic Charter 'but as to the eighth we should learn from experience and seek a negotiated peace and not a dictated settlement'. British policy should assume that the war would end with a German revolution and that a 'workers' government' would be in power. This view, he claimed, was shared by the Socialist parties in those territories now occupied by the Nazis. Dalton replied he was quite wrong to believe this. No one would insist more on German unilateral disarmament

---

[42] There is no record of his having done so.
[43] See above, p. 42.

than these parties. Furthermore, the influence the SPD could hope to exert on British propaganda bore a direct relation to the plans it wished to promote. 'The British would certainly not welcome propaganda that was in conflict with the Atlantic Charter', he declared. Finally he warned Stampfer that what the Labour party could, in any case, do for the SPD was limited by the fact that it was in a coalition government. 'Stampfer accepted this point saying he knew only too much about coalitions and they would never have them in Germany again.'

Stampfer had gained Dalton's promise that he would look into Secret Service use being made of the exiles but apart from this his interview had achieved only one thing, the widening of the gulf between SPD in exile and Labour party. In a last-ditch attempt to save at least something, he issued a statement on 4 December 1941. In it he made only modest claims for the purpose of his visit to London. His primary aim, he said, was to 'rebuild the contacts that the SPD had had with its friends inside occupied Europe'. Secondly, he wished to promote the SPD's ideas on 'German re-education' after the end of the war. Thirdly, he had wanted to launch a 'moral offensive' by radio and leaflet with the help of the British authorities.[44]

This statement could not save Stampfer's visit which ended with his total humiliation. He was cited in a major debate in the House of Lords and he was also thrown out of his lodgings in London in an émigré rumpus which was one of the lighter moments in an otherwise dark story.[45]

Stampfer's visit, the very presence of the SPD in London, the fact that it was so anxious to assist in the formulation of British policy towards Germany, and its keen desire to work closely with British Socialists were factors which were leading to a building up of political pressure inside the Labour party. Part of the leadership of the Labour party was adamant that it did not want the SPD's help but it was not quite sure how

[44] LPA Int. Subcmtee mins. and docs., 1941.

[45] Stampfer had been accommodated at the house of the Czech exile leader Josef Belina. Belina, however, was a friend of Walter Loeb. When Mrs Belina overheard a phone conversation between Stampfer and Vogel which not only lasted half an hour but contained rude remarks about Gillies she physically ejected Stampfer. Stampfer claimed that in reality pressure had been put on Mrs Belina by 'the Gestapo tactics' of Mrs Loeb. See SPD *Mappe* 18 and 73.

to make this more generally acceptable. The dealings that had gone on during the summer and autumn of 1941 proved that what the SPD wanted and what the Labour party wished the SPD to want could only be related to each other with great difficulty and with a large measure of goodwill. The SPD was ready to work at this in the interests of solidarity and Socialist comradeship. It was the Labour party who found such goodwill increasingly difficult to find. Gillies's and Dalton's probing of the inner logic of the SPD's exile existence had demonstrated that there were fundamental differences of opinion between the two sides. As British policy became more Germanophobic these gaps widened.

The tragedy here was that it was not necessary to stress the existence of differences of opinion. Naturally the changes that were taking place in British public opinion towards Germany had to be taken account of by the Labour party. Opinion polls conducted by the government showed the extent of anti-German feeling: in May 1942, for example, thirty-four per cent of the population wished to see a 'preventive solution' of the German problem, twenty-four a 'revengeful' solution, and twenty-three a 'constructive solution'. The respective figures changed radically by October 1942, however, when thirty per cent favoured a 'preventive' solution, thirty-two a 'revengeful' one, only eight per cent a 'constructive' one, and twenty-six per cent did not know which to choose. But by May 1943 the polls were back to their May 1942 findings, until April 1945 when forty-four per cent supported the 'preventive' solution, twenty-eight the 'revengeful' one, and a bare seven per cent a 'constructive' one.[46] Although these figures suggest that support for Stampfer's concepts from British Labour was quite out of the question, support for some sort of co-operation could have been justified. In addition, of course, to a certain and undemonstrable extent, public opinion towards Germany might have become less bitter in the light of greater knowledge of the SPD and its aims. Then again, further work could have been undertaken in secrecy so that longer-term interests would not have been sacrificed to short-term ones.

[46] I am indebted to Professor Michael Balfour for showing me the results of Tom Harrison's polls. See M. Balfour, op. cit.

The Labour party's changed response to the SPD in exile was now in need of official formulation. For this purpose Vogel and Ollenhauer were informed by Gillies on the day after Dalton's interview with Stampfer that the NEC international sub-committee was considering two memoranda on the SPD.[47] The first one was by Gillies himself. It had been circulated since 10 October 1941 and was entitled 'Notes on the SPD during the First World War and on the eve of the Third Reich' and was researched with the aid of Walter Loeb.[48] It was the aim of this document to scotch any further attempt by the SPD to act on British policy through the good offices of the Labour party and it succeeded. It unleashed arguments within the Sub-committee which ended any official contact between SPD and Labour party for the duration of the war.

What took the apparent form of a pseudo-historical debate about the SPD's role after 1871 was, in reality, a crude but effective attempt to destroy the SPD in exile. Gillies set out to show that the SPD had, in fact, always been nationalistic and militaristic and that the SPD's refusal in exile to agree to the policies supported by the British Labour party proved its nature had not even been altered by events after 1933.[49] Gillies claimed that the SPD had said during the First World War that a victory for Germany would be a victory for the working man and 'on the eve of the Third Reich' its policies had been 'equally treacherous'. At first the SPD hoped the Nazis would tolerate them and it was only after Hitler made it plain that he would not do so that the SPD decided to go into exile.[50] Even so, Gillies stated, many of the SPD's *Reichstag* party decided to support Hitler's foreign policy despite a courageous speech by Otto Wels. Although Vogel and Stampfer had also demonstrated a certain bravery, the SPD had supported the Kaiser and had opposed Nazism only because Hitler refused to leave the SPD alone.

[47] SPD *Mappe* 191.

[48] SPD *Mappe* 73.

[49] LPA Int. Subcmtee mins. and docs., 1941. See Burridge, op. cit., pp. 59, 60, 61 and Röder, op. cit., (1969a), p. 151.

[50] See above, p. 13. It was not without irony that the *Vaterlandslose Gesellen*, as the Kaiser termed the Social Democrats, were now seen as guilty of aggressive nationalism.

Philip Noel-Baker, the MP for Derby South, was quick to appreciate that what was at stake was the future of the relationship between the SPD in exile and the Labour party and he insisted that he be enabled to prepare a counter-interpretation of the policies of the SPD. This committee considered his document on 25 November 1941 and, at the same time, various Social Democratic leaders including Loeb and Vogel were invited to make their own submissions. Gillies quite clearly wished the Labour party's attitude to be firmly decided on the basis of historical evidence which he was convinced could lead to the boycotting of the SPD.

Noel-Baker's main thesis was that British Socialists

ought to hope for a revival of German Social Democracy. Our policy ought therefore to encourage now all German Social Democrats both in Germany and outside who are genuine opponents of Nazism and to prepare to give every assistance to a Social Democratic government in Germany when the war is over.

The aims of British Labour and the SPD were similar in many ways, Noel-Baker argued. It was neither anti-German nor anti-SPD to want a 'thorough social revolution', the extirpation of Nazism, the dissolution of the General Staff, the expropriation of the Prussian Junkers, the complete purge of reactionary elements from the civil service, and the nationalization of iron, steel, and other industries. These points were of crucial importance, it should be noted, because they demonstrate the precise basis on which co-operation between the two Socialist groups would have been possible.

Noel-Baker concluded with a harsh attack on Gillies. He could not understand why the notes had been circulated, why Gillies wished to portray the SPD as the 'willing tool of the Kaiser and accomplice of the Nazis had they allowed it to be so'. 'Comrades like Vogel and Stampfer who had risked their lives in the struggle against the Nazis' were being disgracefully treated. In short, Gillies was 'doing Goebbels's work for him' and undermining the credibility of the SPD which was in reality 'a potential ally which deserved the highest consideration'. The facts of the SPD's past had been entirely misrepresented. It was understandable that the SPD should have seen the outbreak of the 1914 war in terms of the self-defence of Germany. Many British Socialists had been equally nationalistic.

The seventy SPD deputies who had voted in favour of Hitler's foreign policy in 1933 had been guilty of a 'terrible mistake' but this was no worse than the errors made by 'Macdonald, Snowden and Thomas in 1931'. It was untrue that a Social Democrat, Schiff, had travelled to the capital cities of Europe as Goering's agent to try to drum up support for Hitler's foreign policy statement in 1933. Thus, Noel-Baker ended, Gillies's views were 'not in accordance with party policy' and he should be reported to the Executive.

Noel-Baker had the full support of Harold Laski. On 19 November 1941 he had written to Noel-Baker saying he found Gillies's views 'malicious' and wanted the National Executive to deal with him. He strongly opposed the 'blackening by perverse and malicious innuendos of the record of men who fought honourably and, at grave personal risk, for the Socialist cause.' On 21 November 1941 Gillies's notes were sent to Vogel, Ollenhauer, Schiff, Geyer, Loeb, and Huysmans. By 26 March 1942 the self-appointed judges of the history of the SPD were ready to pass their verdict.

VI

The SPD and the Labour Party: Dashed Hopes

THE beginning of 1942 saw the SPD in exile fighting to save its political life. Those who wished the SPD to renounce its independence and its belief in the democratic potential of the German people were ravaging the party from within and from without.[1] Leading members of the SPD were preparing to pull out of party activity and the British enemies of German Social Democracy seemed to scent victory. The few friends of the SPD in London were incapable of mounting an effective defence. A book like Harold Laski's *The Germans — Are they human?* had been intended to rally support for fair treatment towards the representatives of anti-Nazi Germans but few appeared to be impressed by this or similar works.[2]

Matters were hardly made more easy for the SPD when a British Sunday newspaper, the *Sunday Dispatch*, decided to make a public issue out of the SPD's role in exile.[3] Under the headline 'the Free German Trick', the paper alleged that the SPD

was pursuing activities tending to counteract the military victory of the allies in order to enable Germany to plunge the world into a new war. Supported by influential British subjects (the SPD) is trying to secure official recognition for a Free German Movement designed to save Germany from the effects of disarmament, occupation and control.

It was catastrophic for the SPD to become associated in the public mind with proposals like these at a time when Britain was fighting its battle for survival. At best this article would

---

[1] On 7 Feb. 1942 Geyer resigned from the Executive. See below, p. 170.

[2] H. Laski, *The Germans — Are They Human?* (London, 1941). See also R. Acland and H. Fraenkel, *The Winning of the Peace* (London, 1942) and H. Fraenkel *Help Us Germans to Beat Hitler* (London, 1941).

[3] *Sunday Dispatch*, 8 Feb. 1942.

lead to public humiliation for the party but at worst, and this was by no means an unreal fear, it could lead to a banning order being placed on the SPD. The British authorities were bound to be harsh on any group which attempted to 'counteract the military victory of the Allies'. The fact that the SPD did not wish to form a Free German Movement and the fact that the Foreign Office had not the least intention of recognizing one was beside the point.

Vogel and Ollenhauer decided to turn to the Labour party for support in refuting the damaging allegations. Jim Middleton the Secretary of the Labour party, one of those who consistently supported the SPD was told by Vogel that it was not only the SPD that was being attacked in the *Sunday Dispatch* but the 'entire British Labour movement'.[4] Most distressing of all, however, was the SPD's fear that its liberties in exile might be curtailed. It was aware of the fact that it was only because its complete loyalty to the Allied cause had never been officially doubted that it was allowed the considerable freedom it enjoyed. Citrine was also approached by the SPD leadership and told a right-wing plot lay at the root of this article and another in the *Daily Mail* two days earlier in which Crossman had been attacked.[5]

Needless to say the SPD's efforts paid no dividend. Middleton did pass the letter on to the NEC but it wrote back to say it was 'unable to help in any way' although it did offer 'sympathy'. When the following Sunday, the *Dispatch* repeated its charges Vogel and Ollenhauer decided to pursue a legal course. They wrote to the solicitors of the paper in order to 'defend the honour of our organisation and its members' to point out the SPD had nothing to do with the Free German Movement, the KPD, or Strasser's Black Front. They pointed out that the SPD had offered its help in the political war against Hitler and that many Social Democrats were serving in the British forces or working in war industries.[6] No reply was ever received and any attempt to start a libel action was, for financial reasons, quite out of the question.

But worse was to come. On 13 February 1942 Gillies was

[4] LPA M. Box 10.
[5] LPA M. Box 10.
[6] SPD *Mappe* 12.

able to write to Dallas the chairman of the NEC International Sub-committee that he had received all the documents necessary for the reaching of a final judgement on the SPD and its political future.[7] The *Sunday Dispatch* affair paled into insignificance. Gillies wrote that his views had been amply confirmed by everything that he had since read. 'I have nothing to extenuate, nothing to withdraw.' He gave evidence of his determination, a determination he wished to be official Labour policy, of either forcing the SPD to conform to Labour's intentions for it or winding up its affairs altogether. The time for such actions was now ripe because the Labour party enjoyed a 'unique position' with regard to the SPD and also to the other European Socialist parties who were similarly beholden unto British Labour for their survival. The Labour party had

An obligation of leadership to fulfil, now and in the future. It should be based on the knowledge, background, spirit and conduct of the leadership of German Labour in historical crises.

Gillies was no enemy of German Socialists, he asserted, for in 1912 he had met Eduard Bernstein 'and become his close friend'. Finally Gillies alluded to the internal debate within the SPD and stated that this was nothing more than the old struggle between Independent and Majority Socialists, a remark which reflects the influence that Loeb and Geyer now had on him.[8]

   Gillies circulated the memoranda on the SPD which he had received from Huysmans, Loeb, Geyer, Schiff, Ollenhauer, Vogel, and Stampfer. Huysmans declared himself

a deceived Socialist. The German nation is infected by the spirit of violence and I am convinced they will follow Hitler obediently and wholeheartedly as long as he is able to speak as a victorious man. He is a national hero.

Previously Huysmans had believed that Socialist ideals could be achieved in Germany but that was an 'error and there is no risk of a further deception'. Loeb was yet another deceived Socialist. He wrote that Germans had insufficient 'civil courage'. One should not forget that a senior SPD member, Löbe, had supported Hitler's foreign policy and that, during

---

[7] LPA Int. Subcmtee mins. and docs., 1942.
[8] LPA Int. Subcmtee mins. and docs., 1942. See below, p. 135.

the First World War, the SPD had been annexationist and
that it had 'collaborated with the Nazis' during the Second.
The third Socialist to see an apparent error in his past ways
was, of course, Curt Geyer, one of the outstanding leaders of
the SPD in exile whose book *Die Partei der Freiheit* published
in 1939 had been an orthodox expression of the Social
Democratic position and a gospel of exile politics. His were
the most radical of the views expressed to the Sub-Committee,
the most persuasive and the least emotional. Unlike Loeb he
did not conceive the problem as a matter of historical morality
but solely as a political one.

Geyer argued that the SPD had always been nationalistic
simply because the German people had always been nationalistic. Noel-Baker was quite wrong to suggest that the SPD
represented those inside the Reich who opposed Hitler. No
one knew what Germans inside Germany believed, that was
the problem. Yet the indications were that they would not
turn to the SPD of their own free will after Hitler's defeat.
The only way in which an SPD-led government could come
about would be if one were imposed on the German people
by the Allied military forces. But in the long term, because
the 'spirit of the masses will remain nationalistic' this would
be 'fatal' to the ideals of Social Democracy. It would not be
able, once democracy had been restored, to 'swim against the
stream'. Thus, Geyer concluded, for straightforward political
reasons, the SPD was a spent force. It either pandered to
German nationalism and ceased to be Social Democratic or it
retained its belief in Socialism and internationalism and ceased
to stand any chance of gaining power. The best thing for the
SPD was to disappear, in exile in London.

Vogel and Ollenhauer on the other hand declared that they
were ready to accept that the SPD's policies during the First
World War had been in part mistaken.[9] But in their defence
they pointed out that the Reich had not been a democracy
and the SPD had not received sufficient information about
the true nature and conduct of the war to act accordingly. It

[9] For the debate about the SPD's role in the First World War see I. A. Barring-
ton Moore Jnr, *Injustice* (London, 1978); C. Schorske, *German Social Democracy
1905–1917* (New York, 1972); A. J. Ryder, *The German Revolution of 1918*
(Cambridge, 1967); and Stampfer, op. cit.

was also a fact that the SPD had made some 'overoptimistic estimates' about the likely duration of the Third Reich and that it had 'harboured certain delusions about the spirit of resistance amongst the bourgeoisie'. Yet all this, and the admitted failings of the Trade Union movement and of individuals like Wels who had been the first to accept he had erred, could not

undermine the historical achievements of the German Labour movement which stood for democracy, social progress at home and peace in international policies. The SPD's leadership had come to London one year ago, placing its confidence in the men and women of Germany and trusting in a renewed SPD. We believed such confidence would be the basis for cooperation with the British Labour party. We are prepared for the tasks which comrade Noel-Baker has outlined and we hope that comrade Gillies's memo has not altered our position.[10]

Viktor Schiff went into greater detail in an attempt to refute Gillies's interpretation of the SPD's history. He begged the Labour party to pursue a Socialist foreign policy, arguing that Gillies's views were 'Conservative'. Would the SPD have spent 1933-9 warning the rest of the world about Hitler if it had not been his opponent? he asked. Gillies had possessed many opportunities in the past to say the SPD 'was a group of fifth columnists' but he had until recently been the epitome of comradeship. 'Now, however, in blind hatred he hits us below the belt', Schiff declared. Vogel and Ollenhauer would have been better advised to go to the USA.

The final evidence before the Sub-Committee consisted of a statement by Stampfer which had little to do with the arguments outlined above since they merely expressed his fury at coming to London, only to be accused of acting for the German Army, and a joint declaration signed by Geyer and his rank and file supporters, Bieligk, Herz, Loeb, Lorenz, and Menne. They all believed that 'aggressive nationalism is the most powerful political force in Germany today'. During the First World War the SPD had been instrumental in keeping the German people behind the war effort and in 1918, because the SPD did not want to revolt against German nationalism, it pursued policies designed to extend Germany's national power. The SPD had always opposed the Versailles settlement.

[10] For Noel-Baker's plans see above, p. 122.

Hitler was not an accident of history, they asserted, but the outcome of the 'biggest mass movement in history'. The German Army was nothing other than the 'political will of the German people'. The SPD leadership in exile, they concluded, had refused to make open declarations of policy and it was guilty of gross deception by suggesting that the 'masses in Germany are waiting for revolution'.

As far as these historical interpretations are concerned, it should suffice to point out that some of them are more valid than others but that none of them actually represents the total truth. The SPD had certainly been responsible for a host of grave political errors yet the fact remained that it alone was prepared to fight for Social Democracy in Germany both before the rise of Hitler and during the Third Reich.[11] Certainly many of the unwise policies pursued by the SPD were pursued for the right reasons and the best of intentions. History aside, however, this debate was a very serious one. First, because it was an internal SPD dispute in which one faction was supported by an enormously powerful and prestigious comrade, the British Labour party. And, second, because historical vilification, whether justified or not, was being used to undermine the political credibility of Democratic Socialism in Germany.

Exile politics could only prove successful if they achieved necessary political legitimation through the preservation of a distinctive Social Democratic identity and the ability to help to plan and formulate policies for post-war German conditions. The second of these had not come to pass and the SPD was now fighting to preserve the first. Nothing could be more damaging for the SPD in exile than for its very claim to represent Social Democracy in Germany to be opposed by fellow party members. Such opposition could mean the end of exile political activity for the SPD, something a number of men, including Gillies himself, were quite happy to see. Their underlying assumption was simply that the Germans, as a people, were incapable of democratic behaviour and that therefore any political ideas based on democracy could never be realized. Democratic Socialism could in the eyes of Gillies and his colleagues, never become a reality in Germany. Socialism could either be imposed from above, against the wishes of the

[11] See above, p. 16.

German people, or it would be destroyed in accordance with their wishes.

Such a view was neither helpful nor realistic. It ignored the fact that from 1933 to that date the will of the German people had not been freely expressed. It ignored the point that even wicked men may learn from their mistakes, and that whatever sympathy there might be for the Nazis in 1942, might well evaporate once defeat and full knowledge of Nazi atrocities became facts of life. Above all, it was not a Socialist view to take and it owed more to moralizing without charity than to a cool analysis of likely political developments after Hitler's end. Chauvinistic vindictiveness was an understandable emotion in 1942 but the only groups likely to benefit from its existence were those who were most antagonistic to Socialism both in Britain and in Europe.

What Gillies and his supporters wanted to create was an SPD which would follow obediently the dictates of the British Labour party and not the alleged wishes of the German people. Once these men had convinced the Labour party that the Germans were simply incapable of democratic behaviour, they believed the SPD would be sufficiently intimidated to force it into line. It was to be given the choice of agreeing to do as the Labour party said or going into liquidation. What they did not realize was that the SPD could not do this and at the same time maintain itself as a political party. The SPD would, on a basis of co-operation, have fallen in with many of Labour's wishes. It had actually wished to do so. But if the Labour party would not allow it to exist as an independent body, those who would profit from a weak and friendless SPD would be the Communists or anti-Socialists on the right.

What was at issue, was not whether the German working class would arrive at Socialism without leadership, but who those leaders should be. In this sense, then, the dispute was complex for, while both sides believed there was nothing intrinsically Social Democratic about the German people, both sides wished to bring about a fundamental transformation of German political life on Socialist lines. There was nothing anti-German about wishing to create a Socialist economic system to destroy militarism, and fight Junkerdom. What caused conflict was whether the SPD, by itself, was to provide the leaders for this process or whether the SPD should be led in this by the British Labour party. The ends were not in dispute. Yet was it merely a question of strategy?

SPD and Labour party could not agree to share the same fundamental view of politics. The Labour party believed that the SPD's leadership could not counteract the wishes of the German people and because the German people were undemocratic, it would be harmful to agree to an independent SPD. The SPD believed that democracy was only possible if the Germans were led towards it by the Social Democrats. The harsh reality was that the Labour party did not trust the SPD because it did not trust the German people and because it believed that democratic politics could only be conducted on the basis of what an electorate actually thought. The SPD, on the other hand, was less concerned with trusting and more concerned to encourage and lead. Marxism was not a consideration for either party, it should be noted, because both of them argued that Socialism could not be assumed to be the inherent expression of working class politics. In the final analysis, and on balance, the SPD's view of its role was wiser than that of the Labour party. Sooner or later the German people would have to be left to their own devices, and it was better that they come to associate democracy and Democratic Socialism with domestic German impulses than with extraneous orders based on military authority. What appeared, then, to be a debate about the past and about historical guilt was in fact a debate about the present and the future of the SPD.

In March 1942 the official Labour party stamp of approval was put on Gillies's view of the SPD and Germany. Nothing had been said or done to make him alter his opinion and it became party policy. The NEC refused to uphold Noel-Baker's and Laski's complaint against Gillies and a motion put forward by James Walker MP stating that 'the document of Mr. Gillies is true in all essentials and the charge contained in Mr. Noel-Baker's document that he is out of line with party policy, is entirely without foundation' was carried with only two votes against. At about the same time James Walker set up the Fight for Freedom Editorial and Publishing Service Ltd., with Huysmans, Loeb, and Geyer on its editorial board. Geyer also became its press secretary.[12] The FFF group aimed to gain as much publicity as possible for its bitter attackes on German political development and it produced a large number of books and pamphlets which all appeared to prove the thesis that Hitler and the Nazis were the true expression of German history.

The SPD in exile now had very few friends in the British Labour movement. As an antidote to FFF notions two Social-ists, Austin Albu and Patrick Gordon Walker formed the Socialist Clarity Group.[13] Yet there was not a great deal it could do apart from offering moral support. The Fabian Society was similarly comradely. On 15 March 1942 the Fabians' International Bureau organized a conference entitled 'After the Nazis'.[14] The SPD was invited and it proved a most useful platform. Although one Labour MP delegate called the Weimar Republic a 'complete sham' and criticized the SPD for destroying the workers' and soldiers' councils in 1918, there was little historical recrimination. Richard Crossman was anxious to deal with practical questions such as, 'Are there forces inside Germany that can be organised from the outside to help us to win this war more quickly?' In addition, this conference gave the SPD the opportunity to say some of the things it might have wished to put to the Labour party more formally, for example that the Germans needed a revolu-tion and that it could only take place with the help of the British.

Other speakers were Wenzel Jaksch and Viktor Schiff who urged the British to adopt a propaganda line towards Germany similar to Stalin's to which Crossman rejoined that it was simply propaganda for the purposes of undermining the German Army.[15]

The Fabians continued to support the SPD and offer it a platform. In December 1942, for example, Vogel, Ollenhauer, Gottfurcht, and Loewenthal were invited to a conference in Oxford and in November they had been asked to a rally to celebrate the Russian revolution. In May 1943 Ollenhauer and Vogel were invited to the May Day celebrations for that

[12] See Röder, op. cit., (1969*a*), p. 157 ff.
[13] See Burridge, op. cit., p. 63.
[14] SPD *Mappe* 36.
[15] See below p. 158.

year and Ollenhauer was invited in May 1944.[16] The Fabians also arranged for SPD leaders to meet various Socialists informally. In 1944 for example they were invited to speak to Léon Blum following his release from Buchenwald concentration camp and, in October 1944, to a farewell gathering of European Socialists.

At the same time Gillies's anti-SPD view was adopted as official Labour party policy, the SPD's exile role came under close scrutiny in the House of Lords. On 18 March 1942 the Upper Chamber debated the extent of the influence of German political exiles.[17] The debate was opened by Viscount Elibank whose anti-German stance was well-known.[18] He wanted to know how many enemy aliens were used in propaganda work and whether their professed anti-Nazi opinions were considered sufficient proof of their being pro-British. He was here making a specific reference to Stampfer and the *Sunday Times* article which had criticized him.[19] He also suggested that Heinrich Fraenkel needed close scrutiny, since he had admitted that he was 'still proud to call himself a German' and had written a 'mischievous book called *Help Us Germans to Beat Hitler*' to prove this attitude was acceptable. The British people should be armed with a knowledge of 'the German character and German history to fight Germany as a whole'.

He was seconded by Lord Vansittart, who made his maiden speech as a peer. He claimed that anti-Nazism was too often equated with 'good-Germanism'. And he stated that he was quoting Nietzsche when he said that the only way to become a good German was to stop being a German and start becoming a good something else. The SPD should certainly not be seen as anything beneficial. It was 'deeply tainted with militarism and expansionism' and had carried the 'spirit of 1914' into Weimar. Indeed, Vansittart continued, Socialist leaders were now opposed to Allied policies. He gave the case of Stampfer who

in the last war was so bellicose, such a warmonger, in fact, that when the Socialist journal *Vorwärts* was temporarily suspended, the militarists allowed it to be revived with Stampfer as editor in chief.

[16] SPD *Mappe* 36.   [17] Hansard House of Lords Debates 1941–2.
[18] Interview with Lord Noel-Baker, 27 Nov. 1977.   [19] See above, p. 116.

Stampfer's ideas were, Vansittart alleged, being allowed to 'infiltrate British policy'. This could be seen in the BBC's refusal to criticize the German Army, although Crossman might be solely responsible for that, he said. Yet the fact was that German atrocities were being committed by the rank and file of the German Army, Vansittart declared, and not, as the BBC had stated by 'specially recruited sections' of what Vansittart deemed 'scum of the earth, namely the German people'. Britain could only get 'one hundred per cent out of people if you tell them the truth and that is that we are fighting the German nation'.

Vansittart was opposed by Lord Noel-Buxton, a Labour peer who stated that it was foolish to believe the BBC used Nazi agents since 'M15 checks them out and vets them very carefully'. Aliens were not allowed to influence policy simply being used as translators and announcers. Goebbels knew that if Lord Haw-Haw had spoken with a German accent, no one would have listened to him. For this reason German exiles were needed in the BBC. He recalled that on 27 November 1941 the Ministry of Labour had paid official tribute to the 'high value of aliens and enemy aliens in many areas of public work'. He was followed by the Earl of Selborne, Dalton's successor as Minister for Economic Warfare. He defended Vansittart's integrity and stated there were thirty-four enemy aliens employed by the BBC 'because of their knowledge of German language and conditions. They do not write scripts or propaganda.' Thirty thousand words of German were broadcast to Germany every day, he continued, but despite this Britain would not be weak in its attitude towards the German people.

There are no kid gloves in the wardrobe of HMG. Kid gloves have been thrown away a long time ago. We are using propaganda and every other weapon in the most ruthless and thorough manner we can.

Although this debate showed that Vansittart's ideas were by no means held by all, he appeared to have successfully established that the SPD was no longer to be seen as in the Allied camp. As Selbourne pointed out, the only exiles that were being utilized were those who had practical value as researchers and speakers. Any political status was explicitly abrogated for those helping the Allied cause. In this way, then, yet another vital opening for SPD influence and co-operation had been officially closed. At the same time the SPD had to face yet another onslaught from Loeb and Geyer. In a booklet entitled *Gollancz in German Wonderland* they tried, by attacking Victor Gollancz, to discredit one of the few leading people still ready to listen to the SPD.

On 2 October 1942 Gillies sent for Ollenhauer, Geyer, and Heine in order to deliver the death blow to the SPD in exile. They were informed that the Labour party would no longer be able to offer them financial support and that they should at once seek other work. Hans Vogel, on account of his advanced age and inability to speak English, would, together with his wife, be supported for as long as possible. Geyer seemed unmoved by this announcement, and Heine claimed this was because he had money of his own from his publishing ventures and from a bank account which he had opened whilst in Paris. Heine believed he could take care of himself. It was Ollenhauer whose position was most gravely threatened. He alone had the energy, the resilience, and the intellectual ability to carry out the organizational and political work which would allow the SPD to survive in exile.[20] Were Ollenhauer forced to seek other work, he would no longer be able to carry out his political duties and the SPD would cease to be able to function. Gillies justified himself by saying that the SPD had been brought to London:

primarily to establish communications with Germany. All such communications pass through my hands and I read them. There are very few. Recently, Ollenhauer collected from two or three friends in Sweden some censored letters and a few reports.[21]

The second task for which the SPD had been funded was to enable the leadership to undertake work 'preparatory to its return'. Severe splits had occurred in the party, however, and this meant that no coherent plans or policies could be forthcoming. Third, the SPD had been requested to produce pieces of journalism. All that had been done here was that a couple of articles had been written for *Left News* and an attack published on Geyer. Gillies believed, therefore, that although Vogel should be supported for humanitarian reasons,

---

[20] See below, pp. 166, 227.
[21] LPA Int. Dept. Corr. Box 2. See above, pp. 31 ff.

Young people like Heine, who is thirty eight and Ollenhauer who is forty one should try to do something for themselves in times like these. Besides, none of our continental comrades in this country with the exception of the Germans claim to have an Executive and it was expressly said by Hugh Dalton that we regard these people as individuals.

Gillies's accusations were unfair. As far as plans and policies for the future were concerned, the SPD's output was good. Indeed Gillies himself wrote some nine months after to Martin Trammel, the Norwegian Labour leader, that the SPD's output in London was

prolific. But it is not my custom to bring their plans before the party because the only pronouncements that have any validity for the future are those which have been made by the governments of the United Nations.[22]

In addition, it was hardly the SPD's fault that co-operation between it and the Labour party had been so infertile. Inside information from the Reich was almost impossible to come by.[23]

On 10 October 1942 Vogel wrote as a last resort to Middleton, the Secretary of the Labour party.[24] He recalled the telegram that Gillies had sent them in June 1940, and how in October 1940 he had promised them reasonable maintenance. Only Ollenhauer and he himself remained to carry on with the party's traditional work, he wrote, now that Geyer had resigned his party mandate and Heine had been engaged by the British Secret Service to work under Crossman at Bletchley Park. It was quite wrong for Gillies to think their work was unimportant. The SPD had four main tasks.

First, the re-building of contacts with our friends in Germany which has become very difficult since there are now only a few channels through the neutral countries. Second, maintaining contact with Social Democratic exiles throughout the world. London is recognised as the official seat of the SPD. Although our exile is numerically not very strong, politically it is the most important. Third, to maintain contact with the Socialist International and fourth, to undertake preparatory work for the future shaping of German politics, in political, social and economic terms.

It was true that the SPD's work had been hampered by Heine's

[22] LPA M. Box 2.
[23] It was even difficult for the Foreign Office. See below, p. 152.
[24] SPD *Mappe* 140.

loss and by the shortage of funds where the SPD was at a grave disadvantage, Vogel claimed, when compared to the KPD which received 'copious funding from Moscow'. Vogel's letter ended with a restatement of the basic reason why support for the SPD in exile ought to have been a primary concern of British Labour. Its present work was assisting the British war effort and its future work would be to help resurrect the German Labour movement. Not to help the SPD now, he declared, was to argue that this resurrection was nothing more than an illusion. Without financial support the SPD would go under.

Middleton asked Gillies for a detailed explanation of his move. He apparently believed that unless the financial situation warranted it, Gillies's action would not be acceptable for moral reasons and he told him that it was not even certain that he would succeed in destroying the SPD since Vogel had already told him they intended to seek alternative sources of funding. The Jewish Labor Committee had already promised $1,000.[25] So all that might be achieved would be that the Labour party would appear vindictive and petty, and ruin the relationship between the SPD of the future and the British Labour movement. Gillies replied that the fact was that there was no more money. The balance of the relief fund out of which the SPD had been financed was £421. Were Ollenhauer to 'go off our books on 1 January 1943' they could afford to look after Vogel for another two years. Indeed, even this might not suffice since no one could know when the war might end and whether Vogel would be able to return to Germany.[26] Gillies then dealt with Ollenhauer himself. He clearly believed that Ollenhauer overestimated his own role in exile and his potential as a future leader. Whether Gillies believed that if Ollenhauer ceased political work, the SPD would die or whether he believed that in order to kill the SPD it was necessary to force Ollenhauer out of party work, is hard to say. The evidence, however, points strongly towards the latter interpretation. Gillies was determined to destroy the SPD in exile and his refusal to fund Ollenhauer was intended as the *coup de grace*.

Gillies complained to Middleton that Ollenhauer had come

[25] Ibid.   [26] Ibid.

to see him to state that his work for the SPD was 'more important than making munitions'. Gillies claimed he had never told
him to make munitions. Ollenhauer then argued that as a
former secretary of the Social Democratic Youth movement
he was a figure of vital significance for the future of the British
and European Labour movements. 'That', said Gillies, 'is a
matter on which opinions certainly differ.' He had expected
Ollenhauer to make a fuss, he added, but he would have to
realize that 'his political work can be done in the evenings, or
at the weekend'. He did not wish to appear vindictive. Indeed,
he was giving Ollenhauer ample advance warning since neither
he nor the Labour party would want to see Ollenhauer having
to ask for public assistance 'which would probably be refused'.
Yet under no circumstances would Gillies make a grant to
Ollenhauer as he had done 'to the French and the Italian
Socialists'. 'I am anxious to prevent such a contingency arising'.
He cited possible Socialist International fury if the Labour
party continued to pay for German Socialists.

In any case, Gillies concluded, the Labour party might
have the 'matter taken out of its hands altogether as Ernest
Bevin had indicated he has powers of compulsion and a new
comb-out of unemployed aliens has begun'. Middleton did
not like this state of affairs and he managed to get Gillies to
agree to pay Ollenhauer until March 1943. Ollenhauer himself realized how difficult matters would be. He had neither
manual skills nor sufficient knowledge of English for a clerical
job.[27] Although he did receive small sums from SPD members
in Sweden and from the Jewish Labor Committee it was only
when the American Secret Service offered him a part-time
OSS post that he was able to face the immediate future with
more equanimity.[28] His job was to compile lists of Nazi officials, and he seems to have worked in an office at the American
Embassy.[29]

After this time Gillies had almost no official contact with
any of the SPD leaders although privately he occasionally
asked after Vogel's well-being. His final words on the SPD
were compiled on 20 July 1944, the very day that the German

[27] SPD *Mappe* 82.
[28] He worked for George Pratt. SPD *Mappe* 82, March 1943.
[29] Röder, op. cit., (1969a), makes no mention of this work.

opposition to Hitler made its most dramatic move. Gillies's report, compiled for the Labour party, shows his punitive spirit. The SPD had disappeared in 1933, he stated, whether one believed it had been defeated, or it had collapsed, or simply that it was absorbed in to the 'national renaissance' of Nazism. Whether it would reappear was a matter of speculation, as was its leadership, and whether it would 'undertake a re-examination of its history and traditions and acknowledge the political and moral responsibility of Germany for the world catastrophe' was doubtful to say the least. Unless great care was taken, he went on, the SPD might reappear as a 'artificial façade designed to save Germany from the inevitable consequences of barbarism, aggression and defeat'. One thing was certain, however, and that was that the International Socialist community would not accept into its midst an SPD which did not 'accept the conditions of peace and in particular the unilateral disarmament of Germany'.[30]

Not every British Socialist supported Gillies's view. Harold Laski was one of a small number who attempted to amend Labour policy. He took the matter up with the Leader, Attlee, and despite a serious mental illness in 1943 did not let it rest.[31] Attlee begged him and the party not to have 'any more great debates about Germany for the time being' but Laski persisted in informing him that Gillies had simply ignored the vital distinction between political behaviour in a democracy and in a dictatorship.[32] Laski believed the Germans were responsible for not having made a revolution. This was needed, but they were not wholly to blame for this failing: 'no small part of the responsibility rests on the capitalist system in decay'. Gillies decided to answer him and Attlee with a statement about the German character and the Labour party's views on it which Laski had no right to challenge.[33] All that

[30] SPD *Mappe* 68. This was a distortion of the SPD's position. See below, pp. 167, 168.
[31] Attlee papers, Oxford.
[32] See Burridge, op. cit., pp. 107 ff., and Attlee papers.
[33] Gillies prepared a long statement for the 1943 party conference on the Labour party's policy towards Germany which he sent to Attlee. On 30 June 1939 the National Council of Labour had sent the German people a message stating they were Germany's friends. In Nov. 1939 Attlee had offered an 'honourable peace'; on 8 Feb. 1940 the Labour party had issued a peace declaration, expressing

Laski could do was write to Vogel, asking him not to take 'the tragic misunderstandings of a few angry men and prejudiced people as typical of the British Labour movement'.[34] Laski was supported by Noel-Baker who also wrote to Attlee.[35] He complained that party policy was being dictated by foolish bureaucratic procedures. In May 1942, he alleged, the constituency parties had opposed the FFF ideas but the Trade Unions had used their block votes to negate this opposition, thanks to the NUM and T & GWU, but in September 1943, the TUC had voted down an FFF motion.

The result is doubly ridiculous for the party. At conference it was committed by Trade Union votes to a policy which violates its principles and which we now know to be repugnant to the majority of the rank and file of both sections because three months later this policy is repudiated by those who put it through.

The real person at fault, Noel-Baker suggested, was Gillies who had spent more time trying to destroy the SPD in exile

the hope that the German people would 'shorten the war'. Lord Snell expressed the 'general view' when he stated that the 'responsibility of the German people is very great' because every people is in the long run responsible for its government. On 22 July 1942 the NEC and the TUC General Council condemned the 'organised and bestial atrocities in Poland and Czechoslovakia which cast . . . dishonour on the nation which has acquiesced in them' and in 1942 the NEC declared that the 'methodical and organised massacre of the Jews was an unparalleled and stupendous act of barbarism which would always be associated with the name of modern Germany'. If the Germans were not held responsible for the actions of their government, they would be invited to deny their duties and obligations and it would absolve them from any necessity for that self-examination of conscience and moral reformation which should inspire a social and political revolution in Germany. It is a principle which can be interpreted by the Germans to mean that by a mere change of government overnight they may escape the consequence of their defeat. It was for this reason that the Labour party had included in its fundamental principles the portion of the Atlantic Charter that spoke of the punishment of those responsible for the barbarous acts. Labour policy on Germany had to be clear. There should be no possibility of rearmament and permanent guarantees for this. The social and economic relationships between military castes and economic privilege should be destroyed. 'It is also an indispensible condition of the untrammeled growth of democratic institutions that there exist the free association of working men in trade unions for the protection of their interests and the defence of peace, independently of governments. This should lead to the eventual emergence of a Germany governed by a political system whose aims and needs runs parallel to ours' 1943 Attlee papers.

[34] SPD *Mappe* 71, Sept. 1942.
[35] Attlee papers, 13 Sept. 1943.

than carrying out his proper tasks. Had he taken a different view or if the Labour party had possessed

a strong man working for the joint action of all the Socialist movements represented in London, we might have enormously improved the prestige and the prospects of the Social Democratic movements in Europe. We have lost a great opportunity and I suggest the matter ought not to be left where it is.

Two days later, Noel-Baker wrote to Attlee again. He urged a 'complete renewal of Labour policy towards the Social Democratic parties of the enemy countries to encourage them all, including Germany'.[36] Second, he suggested that in order to make this policy a reality 'we shall have to get rid of Gillies and clean up the whole International Sub-Committee'.[37] Neither proposal was successful; indeed Attlee appears not to have answered Noel-Baker at all. Gillies remained in office until January 1946 when he was replaced by Major Denis Healey and his formulation of Labour's policy towards the SPD in exile, remained the accepted statement of the British Labour party's concerns and intentions.

Apart from the occasional word, then, all official communication between the British Labour party and the SPD in exile ceased in the middle of 1943. Vogel continued to be supported by the Labour party and when, in the late summer of 1945, he needed an operation for the illness which nevertheless killed him on 6 October 1945 (the precise time at which Schumacher and Ollenhauer were revitalizing the SPD), the Labour party generously paid all his expenses. Morgan Phillips, the new Secretary of the Labour party and John Hynd, the Cabinet Minister with special responsibility for Germany, both came to his funeral. This act was appreciated, even if it was ironic that it was only in death that International Socialist comradeship could be openly demonstrated.[38] Otherwise it is not surprising to find that the Labour party's policy generated extreme bitterness in the ranks of the SPD, although it is worth noting that there was one feature of the British Labour party which the SPD leadership found very intriguing, and

[36] Attlee papers, 15 Sept. 1943.
[37] See Burridge, op. cit., p. 153.
[38] SPD *Mappe* 14.

that was the extent of internal dissension that the leadership permitted. Ollenhauer wrote

The criticism of the Labour party's policies and leaders by the membership is absolutely amazing. The party accepts this dissent and precisely because of it, retains its organisational unity. Just think of the expulsions we would have made if it had been the SPD. We must learn from this.[39]

The balance sheet of the relationship between the SPD in exile and the British labour party makes depressing reading. There was no way in which German Social Democrats—who had been fighting the Nazis for longer than anyone else—could use their undoubted political talents to assist in the formulation of Labour policy towards Germany and the German people. Individual SPD members made a personal contribution wherever they could. Ollenhauer's work for the OSS and Heine's work for the British equivalent are but two examples of this. E. F. Schumacher, the celebrated economist, who helped the SPD draw up economic plans for the post-war period, was asked by Lord Beveridge to co-operate in the making of the Beveridge plan.[40]

What is so disturbing about the lack of co-operation between SPD in exile and the British Labour movement, is that the ultimate aims of both were so similar. Even Gillies argued that the only way democratic conditions could be reproduced in Germany was by nurturing the growth of organized Labour, although unlike the SPD leadership he placed far more emphasis on its gradual growth and on its trade union dimension rather than on its overtly political side.[41] The argument that both SPD leadership and the British Labour party came out of this conflict intact and that therefore no harm was done cannot be accepted. Serious distrust of each other's motives clouded many political decisions in the post-war period. Even Ivone Kirkpatrick felt constrained to write about the post-war British Foreign Secretary:

one thing which left its mark upon him was the bellicose attitude of the German Social Democrats in 1914 for which he never forgave them. He felt betrayed and it made him more anti-German than anything else the Germans ever did. His post-war experience confirmed his gloomy view

[39] SPD *Mappe* 140, 6 Mar. 1943.
[40] See J. Harris, *William Beveridge, a biography* (Oxford, 1977), p. 435.
[41] See above, p. 94.

of German Social Democracy and he found Dr. Adenauer much less difficult than Dr. Schumacher . . .'[42]

This is a striking condemnation of what happened between the Labour party and the SPD from 1941-6. At best it implies that nothing exiled Social Democrats could say could encourage the leaders of British Labour to accept the seriousness of their policies and plans for the future; at worst it shows that ignorance was accepted and allowed to go unchallenged, and even that it could become the basis of policy. SPD leaders who visited London in 1947 still recalled many years later how Bevin had refused to get up when they entered his office or to shake their hands.[43] It seems reasonable to think that the permanent interests of Social Democracy in Europe required a better relationship between Labour party and SPD than one preconditioned by simplistic historical interpretation or plain hatred.

Why were Labour party and SPD incapable of building on their conceptions of the present and the future? Why did both groups make the SPD's political credibility seem dependent on the very thing that lay beyond its powers of control, namely the past? The Labour party as a whole was evidently so revolted by the experience of the Second World War that it did not wish to grant the German people any political independence.[44] The SPD, on the other hand, did not believe it could exist as a party without such independence. As a consequence the existence of a reasonably respectable Social Democratic history became the touchstone by which the SPD's right to independent action could be judged.

Had they been successful, the SPD's exiled leadership would have worked together with the Labour party to formulate plans to reform German political life in a practical way and to turn Germany from an authoritarian racialist regime which

[42] Kirkpatrick, op. cit., p. 205. In an interview Sir Frank Roberts confirmed that Bevin greatly disliked the 'bloody Germans' as he was wont to call them, although did a lot to assist the German population after 1945; interview 28 June 1979.

[43] The SPD leaders who came to Britain in 1947 on 29 November for a ten day visit were Kurt Schumacher, Fritz Heine, Franz Neumann, and Dr Agartz; Labour party Annual Report for 1947, p. 17.

[44] Leaders like Laski argued that it was precisely because Germany's past had been so tortured that the SPD should be helped to make a revolution in Germany.

used terror and injustice, with complete disregard for individual well-being, into a progressive Social Democratic state, founded on constitutionality and respect for human rights. Four years was scarcely sufficient time to begin to grapple with these issues. That these four years were wasted was a political tragedy. It should not be thought that co-operation between SPD and Labour party would have proved awkward for the SPD after the end of the war, as has been sometimes supposed. Although Willy Brandt was caused political embarrassment by his exile experiences when it was claimed he had fought the German Army in Norwegian uniform, the only people who would have opposed the SPD because of its co-operation with British Labour during the war were those who would not have supported it in any case.[45] There is no record that either Heine or Ollenhauer suffered because of their Secret Service work or that any other exile who worked for a British institution was placed at a disadvantage. Mutual help for mutual advantage would have been hard to vilify.

[45] See W. Brandt, *In Exile* (London, 1971) and D. Binder, *The Other German* (Washington, 1975). Far too little was known about the attitude of the German people to their Nazi rulers to make any certain statements about their view of those who had supported the Western Allies.

VII

The SPD and the Foreign Office 1941–1943

AT the same time as the SPD in exile had established itself in London, the Foreign Office attitude towards German political exiles in general and the SPD in particular was changing. By the end of 1942 that change became decisive so that there was no longer any chance of effective co-operation between the leadership of the SPD and the Foreign Office. Thus the SPD leaders' first period in London coincided with a process which they could not have foreseen and which both precluded serious political work and, ultimately, threatened the exiled party with political extinction. Had the Social Democrats arrived in England in 1939 or in 1940, their story might have been rather different.

There were a number of reasons why this change came about. The Central Department of the Foreign Office had a great many other concerns at this time which seemed more pressing than the future of German domestic politics. Slowly but surely British policy was adapted to the views of those who preferred a straightforward military solution to the conflict. The new Foreign Secretary Anthony Eden wholly supported the Churchillian motto that 'nothing about Germany should be fixed' in order that British hands should not be tied after the war.[1] And the unparalleled success of the Red Army in surviving the Nazi onslaught of the winter of 1941/2 meant that due regard now needed to be paid to the concerns and demands of the foreign policy of the USSR.

Yet it is possible to question the political wisdom of the policy of non-cooperation. Indeed it was questioned within the Foreign Office itself, though never with success. The SPD was depressed by its failure to make any impact on British

[1] Woodward, op. cit. (1976), p. 63.

attitudes towards Germany. Above all the adoption of a 'non-policy' towards the political opposition to Hitler both in the present and in the future by the Foreign Office did not serve British interests as well as they might have been served by a more sensitive approach. What started off as a move to keep options open and retain the initiative for Britain ended with the loss of that initiative and the closing of options both in domestic German terms where British influence on the SPD became practically non-existent and in foreign political terms where some of Britain's war-time allies were enabled to imprint their own solutions on the German people. A 'non-policy' can turn out to be even more restrictive than a real policy.

The encouragement of the SPD in exile by the Foreign Office and the advancement of its cause could have produced benefits for Britain. These would have consisted of assistance in the British propaganda offensive against Germany which might have shortened the war, the utilization of Social Democratic skills in considering post-war reconstruction; not to mention the usefulness of laying down foundations for future political co-operation. It is of course true that such a policy might have caused serious difficulties. The British public would have disliked it and the Russian government would have been most suspicious of it. The governments in exile of the occupied nations would have been agitated. But it is, at any rate, arguable that these and other risks were well worth taking. Simply to extend Foreign Office knowledge of German Social Democracy would in itself have been valuable.

Twice in 1944 General Eisenhower asked President Roosevelt to persuade the British authorities to change their attitude towards the German people. It so happens that what he wanted was also what the SPD wanted, namely for the British to state that they were not fighting the war in order to destroy Germany or the German people.[2] To have co-operated with the SPD, then, would have had strategic importance even if some might doubt the subsequent political advantages. It was feasible to accept some of the arguments the SPD put forward without in any way allowing the SPD to dominate British thinking.

The evidence concerning the non-cooperation of SPD and

[2] Woodward, op. cit. (1962), pp. 482, 483.

Foreign Office suggests that British views were sometimes muddled and sometimes less sensible than the possible alternatives. Had the SPD exerted influence it may well be that a number of these policies would have none the less been retained. But others might have been changed. The blanket bombing of German cities, for example, might have been seen as counterproductive, the refusal to encourage the German people to overthrow Hitler might have been modified. The Foreign Office's insistence that it was pointless to work with the SPD in exile because its leaders would have no political future in post-war Germany was yet a further error of judgement.[3] Clearly not everything that was worth exploring was actually worth adopting. But it is at least right to ask whether ignoring the SPD was the most sensible thing for the Foreign Office to do.

After 1942 it was held that Germany needed to be totally defeated before any deep thought could be given to the future transformation of Germany into a stable western-orientated democracy. The British interest in dismemberment can be seen as a means of deferring actual decisions about internal political matters. Stalin's quite different approach to the same issue shows how effective forward planning could be.[4] Indeed, evidence of Russian ambitions towards central, and even parts of western, Europe are primary grounds for questioning British policy at this time.

Three instances show how Foreign Office attitudes towards German political exiles had altered after 1941 compared with the period before it. On 26 March 1941 a parliamentary question was due to be put to Eden by Commander King-Hall. He wanted to know whether he would agree to set up a committee to organize German political refugees so they could 'assist us in the extirpation of the Nazi regime'.[5] Although the Foreign Office had supported such a plan earlier on, Frank Roberts now argued against it:

We cannot commit outselves to this . . . a large proportion of German émigrés in Britain and America have lost all contact with or desire to

[3] See below, pp. 151, 163.
[4] See a Communist view of this argument: D. Lange 'Der Faschistische Überfall auf die Sowjetunion and die Haltung emigrierter deutscher Sozialdemokratischer Führer' in *Zeitschrift fuer Geschichtswissenschaften* (DDR, XIV (1966), 542–67.
[5] See above, p. 57.

return to Germany, unlike the Free French. If the British government offered encouragement it would have to commit itself to pushing the political fortunes of the émigrés. This would prove most undesirable if, as is possible, the few who wished to return to Germany proved to have no following in that country.[6]

Roger Makins concurred. He said that none of the refugees in Britain were 'men of standing, with the exception of Rauschning', nor were they 'men of substance'. William Strang approved these views and added '*we* cannot *create* such a movement. The first impulse must come from the Germans and of this there is no sign.' Similarly, when the ubiquitous Otto Strasser announced in South America in May 1941 that he was setting up a 'Free German Movement' over there, Roberts minuted, 'Strasser may prove useful to us but there can be no question of the recognition of his or any other Free German Movement as long as they are unrepresentative and out of contact with Germans at home'.[7]

Thirdly, the proposal made by Lord Davies which was still on offer and had been expected to appeal to the Foreign Office was now frowned upon.[8] At first Roberts had liked Davies's idea, believing that it might generate some useful propaganda policies. 'It is hard', he noted, 'to envisage a sudden German collapse unless we can offer some attractive post-war prospect to the Germans'. But Strang disagreed:

we cannot think of committing ourselves to set up in Germany any particular kind of regime. If we could, it would still be foolish to hope that any group of refugees now here, after whatever tutoring, could take over. Thus HMG cannot possibly associate themselves with any school for the future rulers of Germany.

Finally he pointed out that if British policy required the dismemberment of Germany, exiles might offer some constructive advice. Were this the case, however, the contacts the Labour party had with these exiles were sufficient. The Foreign Office did not need to get more closely involved.[9]

Roberts was clearly being asked to change his views and he obliged. What Davies had suggested, he now realized, was 'very high policy'. To create a 'united German government' out of

[6] FO 371.26559 c 2951.
[7] FO 371.26559 c 4523, 1 May 1941.   [8] See above, p. 82.
[9] FO 371.26559 c 7108. Contacts with Labour party did not remain sufficient. See above, Ch. 6.

the refugees in Britain would be difficult and these men 'would be very unlikely to provide the leaders in post-war Germany'. Nor should it be seen as wise to 'insist upon a democratic regime because it showed itself obviously unsuited to the German temperament before the Nazi rise to power'.

The Foreign Office views of German political exiles was changing. It was, perhaps, quite reasonable to dislike the notion of a 'shadow German government'. On the other hand some sort of a Free German Movement, even if only semi-official, might have proved an experiment worth trying. It was quite wrong to think the SPD in exile was unrepresentative either of Germans inside the Reich or of political refugees. SPD leaders were genuinely anxious to co-operate in the defeat of Hitler and it must have been obvious to anyone with the slightest understanding of the Labour movement that one day another German Socialist party would emerge.

Evidence of altered British policy can also be found in the discussion surrounding the German language newspaper, *Die Zeitung*, which the Foreign Office had previously supported strongly.[10] In April 1941 an official had commended the first twelve issues to his superior, remarking 'its daily editorial is quite interesting and harmless from our point of view', and had urged the Foreign Office to continue to support it because

it assists in the process of unification in this country and perhaps in Germany. It is part of (our general propaganda line) that we must admit the existence of two Germanies (a 'good' one and a Nazi one) whatever we find necessary to do to Germans after the war.[11]

By November, however, the next time the paper was reviewed, it was minuted that 'this paper should be abolished. It is ridiculous that we spend good British money to make a handful of German refugees and housemaids contented.'[12] Brendan Bracken himself disliked it and was 'strongly opposed' to its continuation. Although *Die Zeitung* remained alive, it was reduced to weekly appearances.

On 30 May 1941 Eden asked the Foreign Office to review British policy towards those Germans in exile who claimed to

---

[10] See above, p. 60.
[11] FO 371.26554 c 1930.
[12] FO 371.26554 c 12269, 6 Nov. 1941.

represent anti-Nazi policies. It appears to have been Eden's first attempt at establishing that the old view no longer applied. Eden had been listening to the BBC's account of the sinking of the *Bismarck* and had been most upset to hear that 'there were such things as good and bad Germans'.[13] The 'confidential paper' which resulted shows the evolving of a negative policy towards groups like the SPD in exile. Although it recognized the real advantages in supporting them and it mentioned the real difficulties facing the German opposition to Hitler, it ultimately concluded that exiles should be ignored.

The document was distributed throughout the diplomatic service in order to inform officials on British policy 'towards Germans and Austrians who profess sympathy to the Allied cause'.[14] It argued that it was legitimate to draw a distinction between Austrians and Germans because the former were victims of the aggression of the latter. Therefore Britain must 'sympathise with Austrians who aim at freeing their country from Nazi domination'. The fact that Hitler was himself an Austrian was not mentioned. Germany, however, was a 'more difficult case'. The paper noted that there was 'considerable controversy' in Britain about the possibility of making a distinction between the German people as a whole and 'their Nazi rulers'. The official British attitude had been stated on 30 April 1940 by Neville Chamberlain, when he stated that Britain had no 'vindictive designs against the Germans as a whole' even though the German people bore 'a share of responsibility for Hitler'.

This remained the official view although certain things were no longer being done in accordance with it. For example 'references are no longer made to Nazi ships but to German ships'. The paper went on to state that it was 'optimistic to expect the German people to dissociate themselves from the Nazi cause in the present circumstances' and that it was not necessary 'to strengthen their support for Herr Hitler by assimilating them completely with their Nazi masters' but at the same time it was vital not to disregard the problems this gave rise to.

When 'formulating policy towards the "other" Germany'

[13] 371.26532 c 5824.
[14] FO 371.26559 c 2951, 18 June 1941.

Britain should not ignore the feelings of the countries 'now suffering from German oppression'. They would rightly be suspicious of any encouragement of German political exiles 'which might be held to imply commitments after the war'. Thus the government had decided 'to make no attempt to promote the formation of Free German Movements'. In short, full support for and co-operation with German political exiles was no longer possible. Further the paper added there were only three leading exiles Brüning, Rauschning, and Strasser. The first showed no desire to play any political role, the second was 'regarded with suspicion by many Germans as a believer in authoritarian principles', and the third is 'condemned as a revolutionary and ex-Nazi'. Finally the paper pointed out that:

Far from looking abroad for leadership and encouragement the German people would in their present mood be likely to regard the leaders of such movements as traitors in enemy pay. We would be forced (by aiding the exiles) to advance the political fortunes of a number of Germans with little or no following in Germany but even to have earned the active distrust of the population.

There were clearly strong arguments present in this paper. And yet some criticism of the ideas it contained is required. It was at best uninformed to suggest that the only three leaders worth any consideration were those to the right of the centre. At worst it showed an anti-Socialist bias.

The resignation of a leading British journalist F. A. Voigt from the BBC's German Service in the summer of 1941 produced further indications of the changes that were under way. Voigt, who had been the Berlin correspondent of the *Manchester Guardian*, had complained in the bitterest terms about the attitude of his colleagues in the Service. Voigt's argument was that British policy towards German propaganda was being inhibited by earlier attitudes which had led to the distinguishing between Nazi and non-Nazi Germans. This had, in turn, produced a 'soft' policy whose keynote had been to encourage Germans inside the Reich to believe Britain wished them no harm. It would be far better, he suggested, to assert British determination to destroy Germany. The corollaries of such a dramatic change were legion and they were too radical for the Foreign Office to accept in their entirety, which is why Voigt's resignation was accepted. The BBC's German Service would have stopped co-operating with individual exiles and it would have become quite ineffective as an instrument of propaganda.

Voigt argued that although there might be 'other' Germans there was 'no other Germany'. Neither freedom nor democracy had any meaning for Germans and so it was wrong for the British to offer any hope of it. Instead, Germans should be told that the Versailles settlement 'had not done the job properly enough'. Links with exiles in Britain should be cut because almost all of them were 'unrepresentative' of German political attitudes. The SPD was singled out for special attack. It had 'no particular claim to German culture, the general standard was higher on the right than on the left'. German 'poets and thinkers' were all right-wing. In general, the differentiation between Germans and Nazis, and the propaganda it had engendered, had been 'cheap, supercilious and silly'. The four hours of daily output had been 'rightly castigated by Hitler on 30 January 1941 as *démodé* and *émigré*'. Worst of all were the programmes for German workers which had been 'conceived in terms of a Germany that does not exist, and never did, despite being reminiscent of Weimar'. The Social Democratic ideas that emanated from a programme called *Vormarsch der Freiheit* were ridiculous, and 'to identify the cause of the British Empire with the cause of a Socialist revolution in Germany can only cause derision'. Finally, Voigt concluded that

the influence of German *émigrés* on our propaganda is almost invariably wrong. With a few exceptions they live in a world of make-believe since they all belong to factions that failed utterly to comprehend Hitler.[15]

Frank Roberts minuted that 'all concerned with German propaganda are now in agreement with some of his more important points', and the 'real enemy was not National Socialism but German militarism which could outlive Hitler'. He went on

the real problem is that HMG now has no policy towards Germany other than defeat. We cannot expect our propaganda to cause any real rift in

[15] FO 371.26532 c 5824, 30 May 1941. See Delmer, op. cit., p. 15 ff. Voigt mentioned two secret radio stations then operating known as BH and DE, previously called freedom stations. DE's task was to 'create in Germany the nucleus of a future opposition and to influence the trend of German revolutionary movements as soon as they begin to show themselves'.

Germany until military defeat and we cannot hope to achieve any useful object by stating now our intentions towards Germany.

The various reactions to Voigt's report all indicate that although Voigt's arguments were accepted, they were considered to go too far to be implemented at that time. Cadogan added his view that he had to confess that '*Black Record* may be correct but I can't see that it is useful propaganda'.[16] He thought it 'not impossible' to persuade the Germans to oppose the Nazi regime and he argued that it was 'highly undesirable to cement the union between the Germans and their *Fuehrer*'. He concluded, 'I personally should like to foster the idea that the simple equation German equals Nazi is not quite so simple as it seems.' The difficulty about Cadogan's approach, however, was that it was very hard to translate into practical policy. As he himself elaborated a little while later, one either argued that *Black Record* was correct and took the risk of making the German people believe they could expect only to be destroyed, or else one attempted to be more generous to them which implied making benevolent promises about their future treatment.[17] *Black Record*, which held that Germans were racially aggressive, that Hitler had given the Germans 'exactly what they wanted', and that severest measures were needed to reform them, certainly was influential.[18] Vansittart's ideas were taken note of. But at this stage, as we may see, it was not official policy. On the other hand, official policy was no longer as clear about Germany's future as it had been.

On 10 July 1941 there was a meeting to consider the direction that British propaganda towards Germany ought to take now that changes were being embarked upon.[19] It was attended by Robert Bruce Lockhart and Ivone Kirkpatrick for the BBC interest, and William Strang and Nigel Law for the Foreign Office and the Ministry of Information. It was noted that the changes in British attitudes towards Germany and German exiles would now have to take account of a fundamentally

---

[16] *Black Record* was the title of a viciously anti-German diatribe written by Lord Vansittart, originally in the form of articles for the *Sunday Times* and published and reprinted four times in Jan. 1941.

[17] See below, p. 155 ff.

[18] Lord R. Vansittart, *Black Record* (London, 1941), p. 18.

[19] FO 371.26553 c 2660.

dramatic new factor: Hitler's invasion of Russia, and the subsequent intentions of the USSR towards Germany. Stalin's
propaganda line was explicitly dealt with because it seemed
so effective. His line was that 'every German defeat is a victory
for the German people' and that the German people and the
Red Army were essentially fighting the same struggle. Just as
British propaganda to Germany was held by all to imply a
statement of British commitments whether positive or negative
towards post-war Germany, so, too, the Foreign Office believed
that the USSR's line suggested a future policy towards Germany. Whereas in 1939 and in 1940 the British could have
matched the Russian line with ease, by 1941 it was far harder.

This meeting noted that Chamberlain had always been concerned to uphold the distinction between Nazis and Germans
but that Eden appeared to hold a different view. At a speech
at the Mansion House on 29 May 1941 he had stated

We must never forget that Germany is the worst master Europe has yet
known . . . Our political and military plans for peace will be designed
to prevent a repetition of Germany's misdeeds.

He apparently believed that without very stern guidance the
German people might repeat Nazi policies even after Hitler's
defeat, that Germans were inherently prone to political and
military misdeeds. This, however, made things very difficult.
Those at the meeting wanted the Foreign Office to be more
sensitive to their views and needs. They argued that the
British aim was 'primarily to destroy the German war machine'
but they added that 'if in so doing we can offer the Germans
an alternative more attractive than Bolshevism we should do
so'. It was imperative to 'offset the more positive line of
Russian propaganda even at the risk of stimulating a division
in this country'.[20]

The Germanophobia that Hitler's barbaric actions were
giving rise to was perfectly comprehensible. Eden, as a politician, was obliged to take note of them. But in the view of
these officials it was not wise to let Germanophobia become
the corner-stone of British policy. The best interests of Britain
may not have been served by officially echoing the quite
proper public feeling of outrage. Stalin, of course, did not
have to worry about public opinion in the same way as Eden

[20] FO 371.26532 c 7782.

and others. Yet his propaganda line did not prevent the soldiers of the Red Army from fighting with the keenest vigour, nor for that matter, did his claims make these soldiers any less harsh in their treatment of the Germans they were to confront.

Sir Alec Cadogan also discussed these matters. He noted that British public opinion was now severely divided on this question

On the one hand there is the so-called sterile policy of *Black Record* and on the other the considerable volume of opinion which favours an attempt to split a mass of the German people from their rulers.

He himself preferred the latter policy but simply for reasons of expediency and 'not of ideology'. What he meant by ideology is uncertain but he believed the latter policy was just more sensible. Yet he accepted that it was difficult to pursue it 'with the entry of the Soviets into the war'. He apparently feared that a 'soft' line towards Germany might be treated with suspicion by the USSR even though its own line was 'soft'. *'Black Record'*, he added, 'can only cement the German people behind their rulers.' Propaganda should be based on simple questions like 'Why must the German soldier be parted from his family for years and hold down populations?' rather than deal with ideological questions.

On 12 July 1941 Cadogan's statement was considered by Eden. He claimed not to have read *Black Record*. He supported Sir A. Cadogan's 'line of questioning'. He also accepted that

We have to try to win the war and for that purpose we must seek to divide the German rulers from the people and also to win the peace and for that purpose we have to seek to educate future generations of Germans. This will be a long process at best and will have to take place under military safeguards for ourselves.

Yet at the same time Eden stated that

I have no confidence in our ability to make decent Europeans of the Germans and I believe that the Nazi system represents the mentality of the great majority of German people . . . therefore the line I took at Mansion House can be followed.[21]

Although Eden was prepared, for propaganda reasons solely,

[21] Woodward argues that Eden was less hostile towards the Germans than is being suggested here. He states that Eden believed a 'democratically-minded' German government was conceivable and often implies Russian policies were far more hostile than those of Britain. See op. cit. (1976), pp. 22, 31, 76, 209, 222.

to continue to try to prise the German people away from the Nazi leadership, he clearly did not believe this could be successful. This was because he disagreed with the view that the Nazis were holding onto political power through the use of terror. He thought the Nazi system corresponded to the political wishes of the Germans. To reform the Germans would be very difficult. There would certainly be no place for German political exiles in this process and, worst of all, Eden plainly did not accept the view that the exiles in general, and the SPD in particular, could in any way claim to represent the democratic aspirations of the German people for they did not possess any.

One ramification of Eden's stern approach was discussed on 25 August 1941 at a 'most secret' meeting between Robert Bruce Lockhart and Ernest Thurtle of the Ministry of Information. They agreed that the new RAF bombing offensive against Germany and the Russian war

provide the opportunity for a propaganda offensive against the German people designed to turn the prevailing listlessness and apathy of the vast majority into active defeatism.[22]

Bombing the German people was now held to be a more effective way of turning them against their rulers than stressing any more hopeful side to the defeat of Hitler. This theory was based on the researches carried out after the bombing of Coventry where it was noted that hatred of the *Luftwaffe* had in time become transferred onto local government officials. RAF crews were to be asked to drop leaflets at the same time as their bombs which would lay the blame for the destruction on local Nazi officials.[23] Bombing was 'good fear propaganda' and could be used to make the population dissatisfied with their rulers.

Even Richard Crossman, who on 9 July 1941 argued that Winston Churchill ought to have expressly 'included the German people' in his promise to offer British aid to anyone who fought Nazism because 'we need a positive propaganda directive to match up the Russians', had, by October 1941, changed his mind.[24]

[22] FO 371.2653 c 7502.
[23] Although bombing of German areas of population brought the war home to the German people as nothing else could, it could be argued that it did little to increase the unpopularity of the Nazis.   [24] FO 371.2653 c 7787.

The short-term (*sic*) Russian propaganda aimed at the German people which aroused such enthusiasm here, has not in fact been at all effective and we have little to gain by changing our long-term line and modelling ourselves on the Russians.[25]

The British should continue to speak of the 'relatively pleasant existence' the German people could have, but there was to be no suggestion that the British were prepared to make a distinction between the German people and their totalitarian rulers, no suggestion that democracy was possible for them. Indeed, Crossman went on to argue that the positive will to resist 'in order to reproduce the conditions of the 1918 revolution' could only be formed by promoting defeatism. Germans were to be made to feel pessimistic about the future and not encouraged to be optimistic, and thus want to get rid of Hitler as soon as possible. Crossman's arguments were considered so interesting that they were sent to Cadogan in Cairo on 10 October 1941 and passed on to the Chiefs of Staff, presumably because Crossman had urged that a western offensive should only be undertaken after morale had collapsed. It is quite obvious that there could be no place for German political exiles in such a scheme. Even if the British had wanted them to participate in reproducing the conditions of 1918, the only basis on which they could have done so would have involved stressing the political alternatives that the 'other' Germany represented. After all, 1918 would not have taken place had the SPD not seemed to offer a better prospect than the regime of the generals.

One early casualty of this kind of thinking was none other than Rauschning. Lord Vansittart had decided on July 1941 to sponsor a visit to America by Rauschning 'for political reasons'. SO1 told the Foreign Office that the idea was very bad. 'R. is an ex-Nazi and hardly the right person to rally the Americans to the British cause.' Strang agreed, stating that 'the outbreak of the German-Soviet war has changed many things and it makes it a good deal less desirable that R. go to the USA'.[26] A committed anti-Bolshevik like Rauschning was no longer a helpful representative of the German opposition to Hitler. When Cadogan wrote to Vansittart explaining why

[25] FO 371.26533 c 11140, 7 Oct. 1941. The USSR did not alter its line.
[26] FO 371.26581 c 7731, 21 July 1941.

they did not wish him to go, Vansittart replied crossly that Rauschning was not 'nearly so harmful as Social Democrats in exile who are doing a lot of harm and whom I am going to counterattack'.[27] The time for the encouragement of German political exiles was over. When Herbert Morrison said on 20 September 1941 that he did not wish to pursue the idea of a 'Free German Brigade' the Foreign Office decided to put his view on file:

It is worth having this opinion of the Minister of Home Security on record since the Foreign Office are often regarded as reactionary in their . . . refusal to encourage German refugees.[28]

On 5 December 1941 a certain Karl Meyer who had been a member of the Roman Catholic Centre party during the Weimar era wrote to Eden asking for official support for his efforts to create a Free German Movement. His group was not confined to any one political group and included Hoeltermann. Eden's response was to ask M16 to investigate him and his associates. The Foreign Office did not want to let anything go unwatched which might become politically embarrassing.[29]

The new direction that British policy towards German political exiles had taken in 1941 was confirmed in 1942. There was now no way in which German political exiles, whether Socialist or not, could co-operate in the formulation of British policy towards Germany and help in defeating Hitler. The fact that the Soviet Union had survived the winter of 1941/2 made the hostility of the Foreign Office towards the exiles seem even more sensible than before. First, it was obviously vital for relations between Britain and the USSR to be good. Encouragement of exiles by the British might have been taken by the Russians as evidence that the British preferred a 'soft' peace to a 'hard' one. Although Stalin argued that the interests of the Red Army and the German people were one and the same, British authorities did not, at this stage, infer anything more from this than Russian espousal of the 'hopeful' line in propaganda. Its logical consequences were not foreseen.[30] Secondly, the fact that the

---

[27] FO 371.26581 c 9397. See below, pp. 193–4.
[28] FO 371.26559 a 10653.   [29] FO 371.26559 c 13458.
[30] When the USSR decided to recognize a Free German Committee, it was the British turn to become suspicious of the USSR. See below, p. 187 ff. Also H. Krisch, *German Politics Under Soviet Administration* (Columbia, 1974).

USSR was still in the war increased the chances of Allied victory and the concomitant prospect that pieces of the Reich might have to be given to the USSR as a reward for having won the war in the East.

On 12 March 1942 the Foreign Office was handed a memorandum from Stalin which made his view on these matters quite plain.[31] He stated that he wished to see an independent Austria, the detachment of the Rhineland from Prussia and its constitution as a separate state, the ceding of East Prussia to Poland, the Sudetenland to Czechoslovakia, and, possibly, an independent Bavaria. Eden minuted in red ink: 'I have no objection to this in principle.' Clearly, to aid non-Nazi Germans in exile would make it very much harder for the British to agree to this kind of territorial rearrangement of the German Reich. Dismemberment, the destruction of German nationhood, could only be carried out in defiance of the claims of 'other' Germans and not together with them. The best way of co-operating with the USSR appeared to be support by the British for the complete destruction of Germany and the management of its domestic politics by the victors. The remark made almost casually by William Strang in July 1941 with reference to Rauschning's visit to the USA was fast becoming a major factor in British policy.

Early in 1942 Cripps, the British Ambassador in Moscow, asked the Foreign Office to describe the British view towards German political exiles because the 'Russians desire to know our attitudes'.[32] On 12 January 1942 a Foreign Office official wrote that

There is no harm in letting the Russians and M. Maisky know that we do not attach much importance to the German émigrés in this country. On the other hand, we need not tell him that they are, generally speaking, though left-wing, anti-Communist.

This interesting minute implied that Stalin need not learn that the strongest exile group was held to be the SPD because he might suspect that the British would consider utilizing them to counteract Communist influence should it appear.

On 16 February Eden sent an official reply to M. Maisky. He pointed out that the vast majority of German refugees

[31] FO 371.30929 c 405.
[32] Ibid.

were 'of course not political émigrés' but Jews. He went on
to tell him about the Social Democratic exiles. Eden not only
supplied Maisky with information that Harrison had wanted
to withhold but also got a number of facts about the SPD
quite wrong. This was doubtless unintentional and the simple
result of the Foreign Office's refusal to have any dealings with
the SPD in exile. He assured the Russians, for example, that
the SPD in exile was incapable of producing policy statements
because the three groups it had subsumed made unanimity
impossible.[33]

Britain's policies towards Germany had, increasingly, to
take account of the USSR's concerns. On 3 January 1942,
for example, Crossman submitted a paper on propaganda to
Germany which unleashed some interesting Foreign Office
reactions. He argued that British ideas needed to be altered
because the USSR was still fighting:

the invincible German army has been checked and Hitler's reputation
for infallibility is in doubt. We have the first indications that not
everything in Germany is harmonious.[34]

Harrison noted on Crossman's paper that there was a further
element of which 'Crossman was still unaware'. This was that

M. Stalin has expressed very definite views on the future of Europe.
No one can tell what will happen in the Spring but it is not unreasonable
to hope that the Russian armies will prevent Hitler's break-through in
the east. Whatever our own contribution, it is likely that it will be the
Russian Army which will break the back of German resistance and we
must therefore now take very serious account of M. Stalin's views.

If Stalin intended to make the Curzon line his eastern frontier
it would be 'very difficult to deny the Poles the rest of East
Prussia'. In addition Eden was shortly to agree to the desir-
ability of a separate Austria and consideration of a separate
Rhineland or Bavaria. British propaganda which had been
until then ostensibly based on the Atlantic Charter, and the
inviolability of national integrity, would have to be altered.[35]

Cadogan agreed that the time had now come to 'stop offering hope and to preach intimidation instead'. Strang, who

[33] FO 371.30929 c 405. See above, p. 66. In early 1942, Roger Makins was
replaced as First Secretary by Frank Roberts and he was replaced by G. Harrison.
[34] FO 371.30928 c 493.
[35] Germany was not specifically excluded from the Atlantic Charter until
15 Jan. 1944.

received this paper, wrote that he agreed and that he had 'always thought propaganda based on intimidation was best'. Eden then made his own views clear. He 'strongly supported the view that the Germans should be told to get rid of Hitler or face utter destruction'. Yet there was far less clarity when it came to precise statements on the shape of domestic politics in a post-Hitlerian situation. Crossman to whom these comments had been passed believed the British ought to try to produce a German government which might comply with British wishes. This, he surmised, would entail 'working for the evolution of a Christian Conservative political movement to include von Papen', which was to say the least, rather a remarkable suggestion. When Cadogan saw this, he wrote that an aim like this was preferable to intimidation, a contradiction which merely confirms Foreign Office ignorance of the domestic dimensions of German politics.

Harrison attempted to précis these views in order to provide Crossman with a Foreign Office view. He stated that the object of British propaganda was 'to increase confusion among the German governing classes'. Yet propaganda should not be framed in such a way as to 'extinguish the hopes of *das wahre Deutschland* that peace could be achieved if Hitler and his Gestapo are removed'. As long as British public propaganda utterances were based on private deliberations as vague and contradictory as these, 'increased confusion amongst the German governing classes' would bound to result.

The fact remained that the pursuit of a negative policy towards German political exiles decreased the number of options open to British policy makers. The Foreign Office had been very disturbed by an article that was circulated at this juncture written by Dr Goebbels and published in *Das Reich* on 7 January 1942.[36] He claimed that the British now wanted to dismember Germany after its defeat and that British intentions were simply to exterminate the German people. Although Goebbels's eagerness to exploit dismemberment suggested it might be wise to offer 'more hope', as Strang and Roberts agreed, in actual fact such a line had now become impossible. Britain's allies demanded that toughness be maintained and in this they were supported by British public

[36] FO 371.30928 c 947.

opinion. Strang pointed out that the USSR was better equipped to deal with such matters 'because the Russians have no internal public opinion' and they could always count on the support of German communists.[37] Strang was, of course, absolutely right to say that the KPD would support Stalin whatever his views on Germany, whereas the SPD would only support western views if they were benevolent towards Germany.

Eden's pronouncements on the subject of Germany and the German people became increasingly harsh. On 8 May 1942 in a speech delivered at Edinburgh he stated that

The longer the German people continue to support and tolerate the regime which is leading them to destruction, the heavier grows their own direct responsibility for it. No one will believe they are capable of working a (political) system based on respect for law and the rights of the individual unless they have taken active steps to rid themselves of the present regime.[38]

And when on 27 May 1942 a well-known Social Democrat in exile in New York, Karl Frank, managed to gain Stafford Cripps's support for a visit to England to be enabled to re-establish contact with Germany, the Foreign Office was furious. 'This is just the sort of thing that might leak out and cause alarm.'[39] When Cripps replied that he had known Frank for many years 'as one of the most determined underground fighters in Germany since Hitler came to power' who had frequently risked his life by returning to the Reich, the Foreign Office explained his contacts might be seen as 'peace feelers' and they were not wanted.

Not everyone supported such harshness. It is interesting to see that General Smuts wrote to the Foreign Office on 10 November to complain about its attitude towards Germany, possibly because of its repercussions on domestic South African politics.[40] His point was 'hope is important'. Harrison wrote back that hope had been offered in the past but the Nazis had been 'only too successful in convincing the Germans that they and the Nazis sink or swim together'. Eden asked Strang for his view on this complaint and he replied that

German morale is more likely to crack under a desperate fear of what is

[37] See below, p. 221.　　　　　　　[38] FO 371.30931 c 4912.
[39] FO 371.30928 c 493.　　　　　　[40] FO 371.30931 c 4912.

going to happen if they do not get rid of Hitler. Eden minuted on this 'I hold this view strongly'. Sir Alec Cadogan, on the other hand, disagreed. He argued that it was a mistake to 'tell the Germans they are all as bad as Hitler . . . Goebbels's line proves this'.

But by December 1942 Crossman was able to write to the Foreign Office that he now believed that the harder line had led to a 'considerable increase in the influence of British propaganda . . . to intensify the fear of defeat into the certainty of defeat', although he noted that the 'most effective fear propaganda is bombing'.[41] This view, it should not be forgotten did not simply lay the foundations of Britain's propaganda line. It also predetermined the British attitude towards the political opposition to Nazism and its representatives in exile. Propaganda and policy were nowhere more closely intertwined than during the Second World War. Thus any political openings for the SPD in exile in London which could have presented themselves, and perhaps ought to have presented themselves, were systematically closed during this period.

When in October 1942 the Foreign Office learnt that there was a chance that German nationals in the USA might no longer be termed 'enemy aliens', the Foreign Office decided to produce a paper summarizing the drift of British policy towards the exiles. Any Free German movement was to be 'strongly deprecated'. There were no reliable indications that the Germans were ready to abandon their faith in Hitler 'or take active steps for their own liberation' and

Far from looking to German movements abroad for leadership and encouragement, they would in their present mood be likely to regard the leaders of such movements as traitors in enemy pay. Whatever government emerges in Germany after the war will consist of Germans who have been through the war in Germany. Any encouragement of anti-Nazi refugees would be sure to arouse the suspicion of those countries now suffering under German oppression.[42]

Both Roberts and Eden approved this document.

Yet however convincing these arguments seemed to be, they should not be accepted without careful scrutiny. As far as the actual winning of the war was concerned, it does seem plausible to believe that the propaganda impact of some

[41]   FO 371.30928 c 12163, 7 Dec. 1942.
[42]   30929 c 10442.

official support for a Free German Movement would have been most helpful. Germans might have been encouraged to resist in the same way that the French were encouraged by the activities of de Gaulle. It would in addition have questioned the legitimacy of the Nazi regime. It should not be forgotten that Hitler once stated that unlike the Communists, it was not necessary to fear the activities of Social Democrats because they were supported by no foreign power.[43] Finally, British support for the SPD in exile could have been a useful counterweight to Soviet support for the KPD, as the events of 1943 proved.

As far as British interests in post-war German political development were concerned, here too the British policy seems less wise when closely analysed. It was perfectly true that the first German government of 1949 was composed of Germans who had been through the war in Germany. Yet it is not clear why, provided that Britain had supported the full spectrum of exile opposition to the Nazis, such support would have been an embarrassment to the post-Hitler politicians of a new Germany. And had Britain supported the SPD in particular this would doubtless have led to the creation of a reservoir of good will and a possible means of combating communist policies in the aftermath of war. It was not necessary to give such support openly and publicly. Quiet comfort would also have paid dividends. For reasons of state, then, a different policy might have been called for. Even the most superficial understanding of the German Labour movement would have suggested that it would re-emerge in a free society and that exiled SPD leaders would have their part to play in its re-emergence, as was in fact the case.

Disinterest, ignorance, and lack of political foresight were partly the cause of this failing. Yet other factors were also important. The Foreign Office, as we have seen, cannot for example be accused of failing to produce a Socialist foreign policy unless the Labour party is similarly taken to task. For although it was a coalition partner in government, the British Labour party did not, apparently, disagree with Eden and his advisors' policies on Germany. Russian suspicions were also a weighty consideration in this period when the full fury of

[43] H. Picker, *Hitlers Tischgespräche* (München, 1968), p. 64.

of the Nazis was concentrated on the destruction of the USSR, and when the future of Britain depended on the Red Army's success. Finally, the physical pressures on the Foreign Office at this time were immense.[44] A small number of people in the Central Department of the Foreign Office had to deal with a huge variety of issues of which many had nothing to do with Germany, let alone with its future political development. In 1941 it was not at all clear to many officials how Britain was to survive and in such circumstances the ideas of a handful of German exiles seemed less than pressing. Public opinion was increasingly anti-German and a policy which refused to take account of this had little chance of coming to fruition.

At the same time, however, the limitations of this state of affairs are noteworthy. Alternatives were possible, as the early days of the war had shown, and were much needed as subsequent events were to confirm. For one thing was fast becoming obvious. The more that the USSR made its territorial ambitions for Germany plain, the greater became the possibility that Britain's interests might be damaged. As William Strang had quite rightly noted, in the German Communists Stalin had potential and effective German domestic political support for whatever course of action he wished to pursue. There was only one possible counterweight and this was the SPD in exile. Yet the harder the British line towards Germany's future and those who claimed to represent it, the less possible the SPD's support for British policies became.

---

[44] This point was made very forcibly by Lord Sherfield in an interview on 8 May 1979.

VIII

The Internal Development of the SPD 1941–1943

ON 3 December 1941 Ollenhauer wrote to Erich Rinner in the USA about the SPD's position:

We have a lot of work to do here now, but it is not particularly pleasant work. Things are not going as well as we had hoped. There is a fundamental difference of opinion in England about the role of Germans and Germany after the war, and a lot of patience and effort is needed in order to ensure we are trusted by the British Labour party. Hitler's policies have generated too much hate. . .[1]

The SPD leaders were having to confront the grim fact that as the war went on, fewer and fewer people in England accepted that a distinction could be made between Nazis and Germans. In the unsubtle logic of politics this meant that Democratic Socialism and parliamentary democracy would have to be imposed on the German people by the Allies and that therefore the SPD was an irrelevant organization. In addition, the view that the Social Democratic leaders were partners in the fight against Hitler was being increasingly questioned. The ideology of the Second World War was changing. It was becoming less of a struggle between conflicting political attitudes and more of a contest between warring nations. As a result, the perception of the SPD had shifted from its original basis. It was no longer an honourable comrade but an enemy by implication.[2]

The exiles' response to this deteriorating situation was to renew their efforts to prove the SPD was fighting with the Allies and to demonstrate that it did have a political future by producing a number of statements and policy documents

---

[1] SPD *Mappe* 79.
[2] See below, p. 237 ff.

as well as by giving other signs of political life, such as rallies. In December 1941 the SPD leadership decided to publish four main declarations. The first was a statement on the Nazi reign of terror in Europe, which Ollenhauer had been asked to compose; the second was a policy statement entitled 'Observations and Principles for Policies immediately after the End of Hitler', the third was a joint declaration with the German Labor Delegation in New York, led by Friedrich Stampfer; and the fourth was a policy statement on the 'international policy of the SPD'. In this way the German Socialists hoped to bolster up their image in London and demonstrate to the British authorities that they were a political force fighting the same war as the Allies and that they deserved to be taken seriously both then and in the future.

In order to achieve a position of influence in Britain Vogel and Ollenhauer were willing to make commitments that many of their fellow SPD members would not like. These involved the serious question of whether the SPD would support Allied plans for the unilateral disarmament of Germany (which it agreed to do) and whether it would abjure all the territorial gains made by the Nazis. The difficulties implicit in such a position were to become increasingly evident.

As Ollenhauer wrote to Rinner,

Sollmann and Brauer will not like our view of peace. We have to hold to the Atlantic Charter which means that we cannot accept any of the territorial acquisitions of the Nazis, and that we shall have to recognise point eight of the charter too, namely the complete disarmament of Germany, even if we do not like it.[3]

As far as the statement on Nazi terror was concerned it could no longer be published in the SPD's news letter because it had been suppressed on account of the paper shortage.[4] It had to be crudely cyclo-styled which must have reduced its impact. It was brought out together with the statement of the international policy of the SPD. The SPD asserted that German Socialists and trade unionists were now more than ever aware of the responsibilities towards the international Socialist community and especially to Socialists in those

[3] SPD *Mappe* 79, 3 Dec. 1941.
[4] The paper was called *Sozialistische Mitteilungen*; see above, p. 29.

countries which Hitler had attacked. They believed that the war had

> its roots in the structure of German society and its economy which have never undergone a thorough democratic revolution. Heavy industry, the landed estates and the army brought Hitler to power.[5]

The terror used by the Nazis throughout Europe since 1938 had been used inside the Reich ever since 1933 and it had thus created a domestic social basis for itself which the SPD was determined to destroy. The SPD pledged 'complete German disarmament and the making good of damage'. This was a 'question of honour' and the 'recognition of the existence of democratic elements in Germany' was the surest way of ensuring the eventual success of this policy.

As far as the international policy of the SPD was concerned, the same points were made about the need for a thorough social political and economic revolution and the SPD pledged itself to 'provide practical proof of the will of the new Germany to cooperate in peace and with loyalty and energy in the work of European reconstruction'. It would not 'recognise any of the conquests or annexations of the Nazi regime and the free Germany of the future would redress the wrongs inflicted by Hitler as a matter of honour'.[6] Although these two documents are neither humble nor subservient they do demonstrate that, on paper at any rate, the charges of hostility towards Allied plans in those areas that were subsequently made against the SPD, were unfair.[7]

The third important text dealt with the SPD's plans for German political, social, and economic reform in the immediate post-Hitler period. The ideas were in large part the preliminary results of the commissions that Ollenhauer had set up earlier that year.[8] They had met on 2, 9, 16, and 23 September 1941 and on 7, 14, and 21 October 1941. There had been general agreement on all the principles adduced which does suggest that not every exile undertaking degenerated into furious squabbling. There was agreement both on narrow matters such as, for example, that no future *Reichsminister* should earn less than a captain of industry, that 'social

---

[5] SPD *Mappe* 179.   [6] Ibid.
[7] See below, p. 187 and above, p. 85, 97.
[8] SPD *Mappe* 179.

justice' should be specifically guaranteed in the new constitution, as well as on broader issues. These included the expression
of the SPD's view of itself as the 'party of the leadership (*sic*)
of the minority of the German people which is determined to
build up a new Germany after the overthrow of the Hitler-
dictatorship'.[9]

The SPD argued that the liberal capitalist economic system
had to be superseded because of its failures. The new Germany
'must be a Socialist one'. It believed there would be a revolution in Germany after the end of the war and the Socialists
must try to lead it so that on defeat the SPD would be able
to present the Allies with a new German government. The first
steps for this had to be worked out by exiled Socialists rather
than by those inside the Reich because the latter simply could
not make any plans. This argument was, of course, doubly
significant because the exiles re-emphasized their mandate to
plan for the future and to develop close relations with the
Allies in the interests of the 'liquidation of Nazism'. The SPD
envisaged two to three years of provisional government because
there could be no parliamentary elections without the 'proper
preparation of the people', so that young Germans and 'others'
would have to wait for some years before being enfranchized
because they had been 'corrupted' by the Nazis.

In addition, the SPD believed that the workers' and soldiers'
councils of 1918 could serve as a useful example for the way
in which Germany might be ruled by the supporters of true
democracy and Socialism.[10] The new Reich government should
be both a legislative and an executive body gaining advice
from a *Reichsrat*, or council, of two hundred members, openly
supported by a trade union movement which could include
the professions. Surprisingly, perhaps, the SPD urged that the
'leadership principle' or *Führerprinzip* should not be immediately abandoned in the trade union movement and that the
unity 'achieved' in the DAF, the monolithic Nazi controlled
German trade union, should not be given up: 'there ought to
be only one trade union organisation'. Finally, the SPD stated
that although there had so far not been a successful German
revolution nevertheless an opposition to the Nazis did exist
inside Germany.

[9] See above, p. 66.   [10] The phrase was *anschauliches Beispiel.*

These statements were intended to dispel any suspicions that might exist in London about the SPD's true motives and intentions. They showed that German Social Democrats in exile were not crude nationalists, that they would thoroughly purge German society, and that they would not repeat the mistakes they had made in 1918. Like their British colleagues, they understood that the German revolution of 1918 had not succeeded in changing the privileged position of the traditional political élites and they accepted that National Socialism would have had a deep impact on many Germans, especially young ones, by the time that Hitler had been destroyed. Whilst their analysis of what had caused the defeat of parliamentary democracy in Germany may not have been the correct one, it was the one that was most widely held in British circles. Yet despite all this, as has been seen, the Labour party was not impressed.

On the question of subsequent domestic German consumption of such proposals, the SPD leaders were probably being as radical as they could. Certain specific commitments were avoided. There was no pledge to make restitution to the Jews, for example, or to punish Nazi criminals.[11] A manifesto issued two years later by SPD members in Sweden on the other hand was rather more outspoken. It demanded a peoples' dictatorship or *Volksherrschaft*.[12] It also promised rigorous state control of the economy, the punishment of war criminals and an 'understanding for the heroic war effort of the USSR'.

It was perhaps ironic that the first reaction to these plans and proposals was to precipitate a crisis within the ranks of the Social Democrats themselves. Geyer's specific reason for resigning his party mandate was his opposition to the plans of December 1941.[13] At a meeting of the SPD's Executive on

[11] See below, p. 173.

[12] SPD *Mappe* 179, March 1943. By that time, of course, far more was known about the extent of Nazi war crimes. The final document issued at this time was concerned with the same issues that Stampfer had discussed with Dalton. See above, p. 116. The SPD and GLD demanded help from the Labour party in broadcasting to Germany: it wished to prepare for the education of the German people after Hitler by producing a stock of literature and information which was to be financed by American sources. Many of the German anti-Nazis in exile, it claimed, were well-known in Germany and should be enabled to address them.

[13] SPD *Mappe* 3 and LPA Int. Subcmtee. mins. and docs., 1942. See above, p. 127.

9 January 1942 he argued that certain changes had been made in the text that had been already agreed upon and that he had not been informed of these changes. Ollenhauer agreed that changes had been made. According to Geyer, Ollenhauer had gone on to suggest that what they should be thinking about was not the present significance of their resolutions but what they would sound like after they had returned to Germany. It was what Germans inside the Reich were going to say that really counted, and, he allegedly added, 'we must always think of later when we are back and how we are going to justify what we said in exile'. Geyer disagreed with him. Their prime duty, he claimed, was to change the face of German politics. This could only be done by altering their own status: 'We need a new party which we must found and lead. We do not want to be simply the trustees of the old party.'[14] At this point, Geyer later asserted, Stampfer entered the room and 'things became quite wild'. Stampfer said a manhunt had been organized against him, the *Sunday Times* had attacked him and 'Mrs. Belina had been subjected to Gestapo tactics by her friend Mrs. Loeb'. The final straw for Stampfer had been Gillies's memorandum which had been designed to destroy the SPD, a move which Geyer now seemed to support. His memo, said Stampfer,

is the basest libel and most infamous slander and Gillies deserves a jolly good punch on the jaw for it. I have seen the exile swamps of Prague and of Paris but the most stinking swamp I have come across is in London. In fact, the whole British Labour party is nothing but a stinking swamp.

As Stampfer correctly saw, Geyer's concerns had changed. They were that the SPD in exile should now transform itself into the active agent of the British Labour party and cease to exert political independence. It was not so much a question of opposition to specific plans and demands, it should be noted, but to the fact that Ollenhauer was determined to keep a weather eye on what the German people were likely to want after Hitler's end, rather than obediently carry out the dictates of the Labour party. Stampfer accused Geyer of wanting to cause a schism in the party. This had been the real origin of the First World War, he declared, and had engendered the

[14] See above, p. 131.

problems which had led to the second. Geyer rejoined that it was Stampfer's policy during the First War that had directly led to the Second, and he then left.

On 7 February 1942 Geyer decided to resign from the party Executive. It was agreed that in order to damage the SPD as little as possible, the official reason should be given as his 'wish to devote himself to historical research' which 'might lead him into making critical observations about the SPD'. He had resigned in an 'atmosphere of friendly understanding and wanting to continue to maintain cordiality'.[15] Geyer's departure was a serious blow to the SPD, even though the leadership managed to disguise both the causes of the resignation and the fighting inside its ranks. Geyer had been one of the leading Executive members, he had been a very important figure during the Weimar era, and his father had been one of the founders of the USPD. His resignation was made even more serious by the fact that Menne, Lorenz, Bieligk, and Herz decided to follow him.[16]

In a letter to Gillies, Geyer once again rehearsed his reasons for leaving the SPD Executive. He believed that 'the views of the victorious powers on Germany should have priority over anyone else's'.[17] Vogel argued that this was to support a second Versailles which would lead to another war. Next Geyer believed that Germany should be unilaterally disarmed whilst Vogel and Ollenhauer both said the German working class would never accept this. Finally, Geyer declared that the SPD was wrong to argue that Hitler was responsible for the war, it was the 'German people' who were morally liable'. They could never be relied upon to make a revolution and the SPD was quite incapable of promoting the cause of revolution for this very reason. The SPD leadership in exile were using terror tactics and Geyer felt forced 'to fight once more the battle that I waged between 1914 and 1918 against nationalism in the German Labour movement'.

Vogel and Ollenhauer did not immediately appreciate the seriousness of Geyer's resignation and they did not foresee

---

[15] SPD *Mappe* 170. Ollenhauer said the same to E. F. Schumacher in a letter of 5 Feb. 1942, SPD *Mappe* 79. 'Geyer resigned in a friendly way.'

[16] LPA Int. Subcmtee. mins. and docs., 1942.

[17] LPA Middleton papers Box 8.

that it was part of a process that was to lead to their exclusion from any practical external political work. At first they attempted to neutralize his opposition by suggesting he was in the pay of Walter Loeb who was acting out of criminal motives to attempt to destroy the SPD. They claimed that Loeb had taken no part in exile politics until 1941, that he had held no high office in the Weimar era, and had probably embezzled state money.[18] Vilification, however, was a tactic that both sides could pursue and it was not a very profitable one, for Loeb was able to argue that his political antecedents were impeccable because he had resigned from all his party offices on account of the SPD's nationalism in 1928 when it had supported the building of a *Panzerkreuzer*. Loeb said he was every bit as politically significant as Hans Gottfurcht for example who was also Jewish and had claimed that Loeb had left Germany simply for racial reasons and not political ones.[19] Vogel and Ollenhauer were quite right to drop this particular line of attack and concentrate instead on a more reasoned defence of the SPD.[20] The fact remained, however, that the SPD was beginning to lose its political credibility and this was bound to have serious impact on its ability to formulate exile policies.

There was one important area where the SPD failed to seize the initiative, an initiative which it could have seized and one which would have helped the party to regain a measure of political authority in Britain. This was in the provision of forthright policies towards the Jewish people whose terrible fate at the hands of the Nazis was beginning to emerge at this time. At the beginning of 1942 the Allied government had agreed at the conference of St James's Palace that the perpetrators of war crimes would be relentlessly punished.[21] It would have been entirely appropriate for the SPD to have supported this at once and, in addition, to have made some promise to make good the awful injustices carried out by the Nazis and possibly to have supported the demand for a Jewish national state. Yet over a month was allowed to elapse after the

[18] SPD *Mappe* 44, 14 Apr. 1942.
[19] SPD *Mappe* 150.
[20] See above, pp. 128–131.
[21] A. J. Nicholls and Sir J. Wheeler-Bennett, *The Semblance of Peace* (London, 1972) p. 32–3.

conference before the SPD leadership even acknowledged that
it was of significance. And the SPD leaders never acknowledged
that Hitler's anti-Jewish measures would place a future German
state in a very special position *vis-à-vis* the Jews.[22]

In November 1943 Vogel was challenged on this matter by
the World Jewish Congress. His answer was that ever since
1934 the SPD asserted that all people were equal irrespective
of race or religion. He possessed reports of 'practical instances
of solidarity with Jews', and he promised that the SPD would
fight discrimination in the future. But, he added rather patro-
nizingly, the success of its efforts would have to depend

on the behaviour of the German Jews themselves. On their return to
Germany after the war they must through their social and political
actions let it be seen that they only support progressive and democratic
forces in Germany.[23]

On 26 April 1944 Vogel wrote to the Jewish Labor Committee in New York in response to a request that the SPD
support the creation of a Jewish state.[24] He tersely replied
that it was not for the SPD to give any opinion on this matter
nor state what might be 'correct compensation for what has
been done to the Jews'. Bearing in mind that this organization
had paid for many SPD members to escape the Nazis, Vogel's
tight-lipped approach was scarcely fitting. Indeed, it is painful
to relate that on 3 March 1942 Vogel had been forced to write
to this group apologizing for an 'oversight' which had led to
the SPD's omitting the Federation's name from a list of the
SPD's benefactors.[25] Privately SPD leaders expressed their
horror of what Vogel called a crusade of extermination, a
Vernichtungsfeldzug against the Jews. But public support was
lacking. Nor was it that ignorance of the true position was to
blame, since Vogel's speeches for 1943 were written on the
back of the reports compiled by the *Bund*, the Polish-Jewish
Socialist organization, detailing Nazi atrocities.[26]

In January 1942 in response to Allied commitments to deal
with Nazi war criminals, E. F. Schumacher complained to
Ollenhauer about the SPD's apparent disinterest. He received
the reply that the aims of Jews and Social Democrats 'were
different'.[27] The most the SPD leadership felt able to say

---

[22] SPD *Mappe* 140, 11 Feb. 1942. [23] Ibid. [24] Ibid.
[25] Ibid. [26] Ibid. [27] SPD *Mappe* 80.

was that German Socialists had been 'the first to suffer from Nazi policies' and when after the massacre of the entire village of Lidice the SPD leaders believed they could no longer remain silent, the statement they did issue made no mention of measures against the Jews and even the Czechs had to take second place to the 'workers of Mannheim' when Nazi misdeeds were raised.[28] It appeared that the SPD felt more able to declare solidarity with the 'Czech freedom fighters who gave the hangman Heydrich his just desserts' than with the non-Socialist German victims of the same mass murderer.

The final unhappy attempt by the SPD in exile to formulate a policy on this question was made in the summer of 1942. Vogel wrote to his comrades that he had heard the Labour party was planning to have an international rally against Nazi terror in Poland and Czechoslovakia. He had been appalled to learn that no German, Austrian, Italian, or Spanish Socialist had been invited. He argued that the Labour party was trying to exclude the Socialist dimension in the fight against Nazism and so he suggested the SPD ought to publicize the speech that he would have made had he been asked to the rally.[29] Vogel failed to realize that he was to a certain extent, the prisoner of his own ideological trap. When the SPD declared, however truthfully, that its fate was analogous to the fate suffered by others under the Nazis, it was only inviting more hostility and offering even more grounds for believing that it was trying to shirk its responsibilities.

Vogel's speech shows how unrealistic his efforts were:

We Socialists from Fascist countries associate fervently and wholeheartedly with the condemnation of the barbarous atrocities committed by German Nazi fascists against the brave peoples of Poland and Czechoslavakia.

He went on to express 'sympathy for all other victims' stating that the SPD knew from 'its own scars' what fascist torture was like. German Socialists were the 'first to fight fascism and the first to suffer from it'.[30] Not surprisingly perhaps the Italian Socialists refused to have anything to do with this statement and so the SPD was forced to adopt another tactic.

[28] SPP *Mappe* 4, 16 June 1942.
[29] SPD *Mappe* 69.
[30] Ibid.

It sent a letter to the Labour party expressing its sadness that it had not been invited to the rally and stating its eagerness 'to see the defeat and destruction of Hitler and the punishment of Nazi criminals'. At the same time, the SPD wished to recall that it was the 'first victim of Nazi brutality and tens of thousands of German Socialists and trade unionists have been murdered and put into concentration camps'. The SPD was pledged to destroy the power of the German aggressors, for once and for all, and to remove their economic base. It was because of this that the SPD 'felt itself in spirit with those on the platform' and regretted the Labour party had not invited them along.

Ollenhauer was privately quite aware of the difficult position the SPD was in when it tried to deal with these matters. On 5 February 1942 he noted that the entry of the USA into the war meant that the eventual defeat of Germany could no longer be seriously questioned yet at the same time the uncertainty of defeat made everyone think very carefully about whether a new Germany could be an equal partner in the community of nations. 'The unimaginable crimes that the Nazis are committing makes this very difficult', he recognized, and he believed the Russians would be allowed to Bolshevize much of Germany as a consequence.[31]

Although there is much evidence to support the view that the SPD leadership handled these matters rather badly, and although the worse Nazi actions became the more the SPD would have to suffer for its failure to deal with them, the position of the SPD was a very difficult one. It seemed hard to combine both a desire to assume political authority after Hitler on a democratic basis as well as support for Allied policies designed to get rid of him. With a little help from the British Labour movement, it might have been possible for the SPD to devote less effort to justifying its own position and rather more to the issue of what the SPD proposed to do about Nazi crimes. Because the British Labour party believed that the German people were responsible for them, and because the SPD could not agree, too much self-justification was inevitable.

As time went on the SPD leadership became increasingly

[31] SPD *Mappe* 80.

reconciled to the fact that a practical and external outlet for policy formulation was not likely to be won. As a result Vogel and Ollenhauer decided to devote the exile to the internal development of plans and policies for post-war Germany. These, it was hoped, could be realized through the inner strength of the SPD. It would have been both better and easier to have co-operated with British authorities but since this was no longer very likely they believed the SPD should go it alone. They knew that the skills of some of the exiles were considerable and that good plans and policies would not only prove the SPD was still alive but would also compensate for their failures in other areas. As things turned out, the SPD met with limited success in this enterprise. Their plans and policies were in most cases first rate. But on the other hand these could not of their own accord guarantee the continued existence of the SPD in exile as a body with its own specific political identity. Rather, as we shall see, it was the German Social Democrats' opposition to Communism that in the event enabled the party to survive.

Consequently, it is not surprising to find that the SPD leaders were most depressed by the direction of events. Their planning did not offer them anything approaching the sort of political work they so keenly sought. As Ollenhauer wrote to Ernst Paul in Sweden on 22 June 1942, they no longer had any money, their opponents Geyer and Loeb received far greater publicity than they did and he and Vogel often thought it would be better if they gave up their party labours altogether and devoted themselves to their gardening.[32] Hans Vogel told his personal friends how miserable their lives had become and how terribly they suffered from isolation and the 'emptiness of exile'.[33] Finally, a new threat could be seen on the horizon. This was the KPD and the fear that as the Red Army began to be increasingly effective in the fight against the Nazis, the fortunes of the KPD would improve in direct proportion. Indeed in a letter to Stampfer, Ollenhauer claimed on 10 July 1942 that German Communists received 'preferential treatment from the British authorities'.[34]

---

[32] Ibid.
[33] SPD *Mappe* 140.
[34] SPD *Mappe* 80.

Nevertheless the plans that the SPD produced must be seen as a courageous attempt to stay in the business of politics even when almost every objective factor militated against this. First, of course, Ollenhauer would no longer be supported by the Labour party and his work for the OSS, which was the only way he could continue to do political work, carried obvious dangers.[35] Secondly, Vogel's inability to master English combined with increasing age and depressive tendencies made the most authoritative figure in exile almost incapable of using his undoubted representative skills. Thirdly many rank and file members were finding it hard to work on party business except on Sundays. Fritz Heine, for example, had changed his name to Fred Harmer, had begun working for Richard Crossman and the British Secret Service, and was shortly to leave for North Africa. But contacts were assiduously maintained and although the fruits of these years were limited in number their quality was high.

As early as 18 November 1942 Social Democrats in exile had agreed to support Ollenhauer's proposal that plans be drawn up for the creation of a single Socialist party in Germany after the war.[36] The term that was used at the time was *Einheitspartei*, a name later used by the Socialist and Communist coalition party which took power in the Russian occupied zone of Germany. Yet there can be no question that in this case the term did not imply any such coalition but simply a continuation of the Union arrangement which had been reached during the first months of exile.[37] A firm commitment to the principles of Social Democracy was to be the basis of party cohesion. Ollenhauer outlined four different areas that needed examination before any final statement about the policies of German Social Democrats in London could be made. These were, first, the organization of such a party and the identification of the political tasks it would have to deal with in the period after the collapse of the Third Reich. Second, they would have to devise an 'action programme' for the SPD's accession to power. Third, the long-term interests of the party would have to be considered and

---

[35] See above, p. 138.
[36] SPD *Mappe* 4.
[37] See above, p. 89 ff.

fourth, the SPD's relations with the KPD would have to be carefully defined.

It has sometimes been alleged that these plans may be seen as the 'export of English democracy' to Germany by the exiles.[38] Although this notion is attractive in broad terms, it is not wholly illuminating. The nature of English democracy is, for a start, not easy to describe. Whilst it was certainly true that some aspects of British politics, most notably the tolerance that was felt to exist within the British Labour party towards a variety of different views, were greatly appreciated, others were not. On 6 March 1944, for example, Vogel wrote to Heinig in Sweden that

English tolerance must also be an important aspect of the post-war German Labour movement. Within the Labour party there is criticism of the leadership which we cannot even conceive of. Yet the Labour party accepts this and maintains a unified organisation. We would have simply expelled members who criticised us.[39]

Yet, as Vogel and Ollenhauer must have realized, such tolerance was not all-embracing. The Labour party was perfectly capable of expelling leading figures like Cripps and Bevan, not to mention D. N. Pritt KC, MP who was expelled from the party because he refused to denounce the Russian invasion of Finland. Where Communists and Communist associations were concerned, expulsion and non-admission were devices that the Labour party did employ and which were wholly consistent with the SPD's attitude to these matters. Tolerance combined with anti-Communism, then, was something common to both parties' views. Tolerance on its own had very little meaning to either. There seems no evidence for Röder's view that, because of a concern for tolerance or otherwise, Ollenhauer was prepared to consider accepting the KPD into a united Socialist party on condition that it left the Comintern.[40]

[38] See W. Rudzio, 'Export englischer Demokratie?', *Vjhzg*, 17 (1969), and W. Röder, 'Deutschlandpläne der Sozialdemokratischen Emigration in Grossbritannien 1942–45', *Vjhzg*, 17 (1969*b*), 72–86.

[39] SPD *Mappe* 14.

[40] Röder, op. cit. (1969*b*), 77. There is some evidence to support the view that what some members of the British Labour party may have wanted was an amalgamation of KPD and SPD to produce a united working class party. In October 1943 Harold Laski wrote to Vogel that the one thing he missed in the SPD's declarations on policy was 'the absence of any pledge to attempt to reach unity in

Thus when Ollenhauer spoke about the 'possibility and tasks for a future single Socialist party' his stressing of 'Socialist reforms' and his Marxist interpretation of Nazism did not imply that he was flirting with the KPD. He argued that German society needed radical transformation if aggressive nationalism was to be defeated, but he always demanded that the SPD should have full 'organisational freedom' and that those who could not give their allegiance to Social Democratic ideals should be excluded from the party. As Ollenhauer wrote to Reinbold in Sweden on 28 March 1943 he did not intend to found a new party in exile and as Vogel assured Rawitzki

there can be no cooperation with the KPD. This would astound and upset our colleagues in America and Sweden. We do not believe that our shaky unity should be jeopardised by even temporary cooperation with Communists.[41]

Indeed, it is surprising how intensely anti-Communist the SPD leaders were prepared to be at a time when considerable pro-Communist sympathy might have been considered quite acceptable in view of the heroic efforts of the USSR.

On 5 March 1943 Ollenhauer submitted his organizational plans for the various committees to advise the SPD on the reborn party. They were to take as their starting point a Germany 'in the territorial borders of Weimar, subject to disarmament but with no constraints on free party work'. There were to be two groups or commissions, of which one was to be chaired by Ollenhauer and Heidorn and the other by Schoettle and Neumann. Commission one, as it was called, was asked to devise an 'action programme'.[42] Its membership included Vogel and Ollenhauer who were to advise on economic and trading policy, Kreyssig and Schoenbach on tax and finance, Rawitzki on the new constitution, and twenty-two other experts on matters ranging from issues concerning women to the supply of foodstuffs. Commission two concerned itself with the organization of the party itself, with Sander and Ollenhauer advising on the central matter of its organization.

the working class ranks which seems to me the precondition for the achievement of a democratic Germany. I also regret the absence of any undertaking to give the Jews equal rights in post-war Germany.' SPD *Mappe* 71, 20 Oc. 1943.

[41] SPD *Mappe* 80, 141.
[42] SPD *Mappe* 4.

Heine contributed a massive document on the reorganization of the German press. Vogel dealt with Germany's international relations in the aftermath of war. All the deliberations were to be completed by 30 July 1943.

These plans are not easy to evaluate, not least because the objective situation that confronted Social Democrats in 1945 was different from the one they had envisaged. Yet the status of the plans is not entirely connected with their prophetic ability but is to be seen more as statements on the SPD's perception of itself during the exile. For these plans represent a very serious attempt to remain in the arena of practical politics in order that, should the occasion present itself, the SPD would be ready to assume political power. They were less concerned with the prediction of the future than to stake out the SPD's claim to power and, as such, are to be seen primarily as a strategy to assume the leadership of the German Labour movement on the fall of National Socialism.

Nowhere does the political seriousness of the exiles' plans come through more clearly than in Fritz Heine's documents on the reconstruction of the SPD's newspaper and printing activities, originally formulated in 1942 but resubmitted to the SPD commission from North Africa where Heine was working for British intelligence.[43] His contribution not only demonstrates his energy and dedication as an individual but also shows the degree of political commitment amongst the SPD's leadership.

Vogel's contribution on the other hand was wider-ranging and therefore perhaps a little less impressive. Heine had systematically discussed almost every detail and possibility that existed, pointing out that he did not know which ones would take place. Vogel, on the other hand, appeared unable to confront the rather obvious fact that it was possible that Germany's foreign policy might be dictated to her by the Allies and so, despite offering a partially correct prediction of German foreign policy aims under Adenauer, he somehow failed to sound convincing about it. This was especially regrettable because the SPD went to some expense to have his contribution (which was entitled 'The international policy of German socialists') printed professionally and distributed

[43] SPD *Mappe* 51.

to various British figures.[44] He promised that the SPD would both eliminate the domestic causes of aggression and also support the 'peasant, the intelligentsia and above all the international labour movement everywhere'. He argued that national sovereignty, as a principle, was no longer compatible with the economic and political condition of Europe and that the SPD would support any move to create a united states of Europe aided by the British commonwealth, the USA, and the USSR.

The first objective of the SPD after the war, he went on, was the 'integration of Germany into the international order'. It would be a 'matter of honour' to make reparation for the 'injustices inflicted on other peoples', and German disarmament was also essential. What was needed to bring all this to fruition was 'a thorough-going reform of German educational structures'. At the same time, he concluded, any attempts to impose conditions on Germany which would lead to underemployment would be disastrous and make the struggle against all 'reactionary forces' inside Germany impossible.

The more general statement that emerged from the various deliberations seems to have been produced in 1944 but it is impossible to be sure of the date. In May 1943 Ollenhauer had written to Ferl that their newly undertaken task had caused them 'to look forward to returning to Germany unlike the comrades in the USA'.[45] On 9 June 1943 Stampfer was told that educational policies were now being planned. On 24 January 1944 Ortloff was informed that the economic programme had just been started, and on 8 February 1944 E. F. Schumacher was asked to take over the chairmanship of the economic committee. The final document must therefore have appeared after that time.

The statement began with a disclaimer: it was *not* a manifesto but simply a collection of 'programmatic principles'. This formulation was important: it underpinned the SPD's exiled leaders' intention not to be seen to lay down the party line in exile but to offer a number of conclusions about the future party's terms of reference. As Ollenhauer wrote to Heinig (who had become the leader of the Swedish SPD group

[44] SPD *Mappe* 184.
[45] SPD *Mappe* 80, 29 May 1943.

in February 1943 replacing Emil Stahl) on 24 May 1944 'it would not have been sensible to have created the impression by our plans that . . . we are replacing the party at home. We are its trustees and the representatives of its interests and ideas.'[46] In this way, the SPD hoped to demonstrate that it was determined to lead Germany after Hitler. The principles began with a discussion of the 'economic aims' of the SPD. There should be freedom from economic 'exploitation' and equality of 'economic opportunity'. In addition the SPD committed itself to full employment. These things were to be achieved by the 'liberation of Germany' from the bonds of private monopolies and by state planning at all levels. The difference in 'relative income' was to be abolished and the state would 'take a key position' in industry for which it would pay compensation. It would, therefore, take into public ownership all banks, insurance businesses, coal and other mineral assets, the building industry and building land, and, of course, big estates.[47]

The document then dealt with the constitutional ideas of the SPD in exile. These included that there should be 'a Socialist Democracy born out of co-determination, *"Mitbestimmung"*, and co-responsibility, *"Mitverantwortung"'*, which would produce 'social justice and peace'. Germany was to be a Republic, a unitarian state with widely decentralized powers and self-administration granted to the '*Länder*'. These it was added, were not to bear any resemblance to those 'existing at present'. The constitution was to guarantee the basic freedoms of personal integrity, safety, and the secret and direct exercise of suffrage. Parliament was to last for four years and be elected on the basis of one man constituencies with 'special measures' for dealing with small parties. The president of the state was to be elected by a *Volksrat* for 5 years and he could be re-elected. The Prime Minister and ministers were to be nominated by the president but had to enjoy the confidence of the *Volksrat*. There was to be a *Staatsrat* to safeguard constitutional procedure. In addition, church was to be separated from state 'to serve the best

[46] Ibid.
[47] SPD *Mappe* 186. See the Labour Manifesto for 1945 for interesting similarities.

interests of society' and there was to be a single national health insurance scheme. Universities and colleges of technology were to enjoy the same status and the right to decide whom to admit. Life tenure was to be guaranteed in them and no political interference tolerated. Finally full reparation was to be made to the victims of Nazism.

These policies show that the SPD in exile wished to make a very clear break with the Weimar system. Germany was finally to become a unitarian state as an antidote to the particularist tendencies of both the Bismarckian and Weimar Reichs. The SPD plainly believed that Hitler had been able to use Bavaria as a powerbase and had also used the competing claims of Reich and *Laender* and the resulting state weakness to his best advantage. It was now determined to enact the constitutional proposals that it had been advocating ever since 1875. Small splinter parties, which the NSDAP had been until 1930, were to be outlawed so making it impossible for radical minority interests to gain support through being represented in parliament. The president was to be elected not by universal suffrage as in the Weimar constitution and legality and parliamentary democracy was to be more rigidly safeguarded than ever before.

But it is not only the firm intention not to repeat the mistakes of the first German Republic that is so striking about these exile plans. For, in addition, a number of ideas about state and society were borrowed from the Labour party and from British ways in general. The provision of a single national health service is one example of this and it obviously owes much to the discussions surrounding the Beveridge report which was being produced at this time.[48]

48 See below, p. 142.

PART III
1943–1945

IX

The British Foreign Office and the National Committee of Free Germany

IN January 1943 President Roosevelt declared that the Allies had adopted the principle of unconditional surrender.[1] Much has been written about this policy and its effects. Here, however, it suffices to note that whatever reservations Churchill or Eden may have had about it, it it no way contradicted the line that the Foreign Office had already embarked upon in respect of Germany's future. Unconditional surrender confirmed an existing set of ideas. The notion that nothing should be done which might prejudice the conditions the victorious Allies would impose on their enemies did not, however, in itself imply that more detailed planning for post-war Germany should be ruled out. Thus the changes in British attitudes towards German political exiles are not wholly explicable in terms of unconditional surrender.

As an overriding principle of Allied war aims, unconditional surrender had many advantages. It minimized the danger of a separate peace, it reassured the representatives of the nations at that time suffering from Nazi tyranny, and it lessened possible chances of inter-Allied disagreements. Yet whether the policy hastened the final victory over Nazism or laid the foundations for a better post-war Europe is another matter. It certainly was disadvantageous from a propaganda point of view, for it seemed to strengthen German resistance to the Allies. And where it was used to avoid taking difficult decisions, it was not wholly conducive to the creation of a more stable world.

Ever since the end of 1942 it seemed that the British Foreign Office had clearly decided that German political exiles were in no way to be permitted to have any influence on the

[1] See Nicholls and Wheeler-Bennett, op. cit.

formulation of British policy towards Germany either during hostilities or in the period thereafter. No distinction was to be made between Nazis and Germans, between those who actively supported Hitler and those who were determined to co-operate in defeating him. There was therefore a considerable amount of consternation when the entire issue of the 'other' Germany was reopened in 1943 with the intelligence that German political exiles in Great Britain and the USA were discussing the formation of Free German committees which might not only interfere with Allied plans for post-war Germany but might also sow discord amongst the Allied nations.

At first the Foreign Office was worried by the notion that a 'soft' peace might have to be made if the exile groups gained sufficient publicity. Thereafter more sinister possibilities appeared to present themselves. First was the prospect that the USSR might attempt to make a separate peace with Nazi Germany. After all, the Second World War had begun with a pact between the two dictators, and the western Allies' failure to form a second front in Europe was bitterly resented by the USSR. Second, any exile grouping which achieved success in Britain might prove difficult to ignore once Hitler had been defeated. British authorities did not wish their hands to be tied by any German exile group. This was not only because they did not want to commit themselves to any political movement in exile but also because they did not want to add to the number of interested parties they would have to deal with at the end of the war. The Free French, the Poles, the Dutch, the Belgians, and the Czechs all required to be taken note of, not to mention the other nations, European and non-European, who had legitimate interests in Germany's fate.

Foreign Office concern was rekindled by a report from J. Wheeler-Bennett who was acting for the Foreign Office in Washington. In April 1943 he informed London that various left-wing groups were attempting to unite, something both new and potentially dangerous:

If anything were to come of these moves, a weapon might be forged of which effective use might be made in support of German nationalist and pan-German ideas and of the policy of 'fair play for Germany at the peace table'.[2]

[2] FO 371.34414 c 8000, 13 July 1943.

The immediate reaction of the Foreign Office was to request MI5, the British Secret Service organization concerned with internal security, to investigate the attitude of left-wing German exile groups in Britain towards the question of uniting. In due course MI5 produced a dossier which was the result of research by a number of different sections. It confirmed that in Britain there now existed a 'Free German League of Culture'.[3] It had started to make efforts to

> obtain recognition as representing German refugees in this country under the direction of German communists who uphold the thesis that the mass of Germans have been coerced by a Nazi clique and that widespread opposition exists inside Germany. Whilst German émigrés in this country have been careful not to give open expression of their views, there are among them who are proponents of the 'good' German thesis and a 'soft' peace.

MI5 pointed out that many Socialist exiles had been made anxious by the recent resolution adopted by the British Labour party.[4] They feared that a military government would be imposed on Germany after the war by the Allies in order to prevent a Social Democratic revolution from occurring. Indeed, the report stated, this was why most left wing groups opposed the Allied plans for occupation. Their aim was a peace which would produce a strong Social Democratic Reich with a 'dominating position in Central Europe', which they wished to see as a large Federation. But there was, MI5 concluded, little evidence of any organized propaganda campaign to this end.

The official in the Central Department with immediate responsibility for these matters, Geoffrey Harrison, took this report seriously. He argued that two points were worth noting. First, that German exiles in London were 'very much on their good behaviour and less outspoken than their colleagues in the USA', and second, that whatever their views, they had in fact been able to gain very little influence on any action of British public life 'not excluding the Labour party with which their contacts are strongest'. Harrison was relieved to note that the British were more fortunate than the Americans in that 'there is no German refugee in this country of the calibre of Karl Frank or Brüning or Rauschning'.

[3] Ibid.; Richard Loewenthal, an SPD member, was allegedly a leader of the League, FO 371.34416 c. 15186, 20 Dec. 1943.
[4] See Ch. 5.

But on 12 and 13 July 1943 a new dimension was added to the situation. A National Committee of Free Germans was formed in Moscow with obvious official Soviet support. It appeared to be a coalition of Communists, Social Democrats, bourgeois, and *Wehrmacht* opponents of Hitler. Its president was Erich Weinert, a poet of communist convictions, and Walter Ulbricht, Wilhelm Pieck, and Colonel von Einsiedel, a nephew of Otto von Bismarck, were among its members.[5] The fact that it had been officially recognized by the Government of the USSR seemed to suggest that it was the first step in a most dramatic direction. Stalin's keen interest in and gift for political propaganda has already been discussed.[6] His celebrated Order of the Day delivered in February 1942 was seen an an expression of this. In it he had declared that 'Hitlers may come and Hitlers may go, but the German people go on'. Official recognition of a shadow government in exile, on the other hand, was something rather different.

The first notification of this development came on the 25 July 1943 when the British Embassy in Moscow informed the Foreign Office that a Free German Committee had been formed there.[7] The Soviet government had clearly not wanted to inform the British authorities that this was going to happen nor did they offer any explanation for it at this stage. It appeared that the Foreign Office had a crisis on its hands. 'What does all this mean?' demanded Eden, the British Foreign Secretary, and he ordered Clark Kerr the British Ambassador to 'ask Molotov'. Attached to the dramatic news from Moscow was a report of the NKFD manifesto. The NKFD claimed that Germany had clearly lost the war, that the patriotic element must overthrow Hitler, and, significantly, create a government to negotiate on the basis of a withdrawal of forces and a renunciation of all conquests. It promised an amnesty for Nazis who joined their movement and a full democratic programme for the new Germany.

On 27 July Clark Kerr telegraphed the war cabinet in London 'with particular secrecy'. He pleaded his 'complete surprise' at

---

[5] See E. Weinert, *Das Nationalkomitee 'Freies Deutschland'* (East Berlin, 1957) and H. Duhnke *Die KPD von 1933 bis 1945* (Köln, 1972) pp. 368-82.

[6] See above, p. 158 and elsewhere.

[7] FO 371.34414 c 8488, 25 July 1943.

these events and said that the US Ambassador was 'baffled'. Various explanations for the formation of the NKFD were put forward by the Foreign Office. First, there was the view that it was simply a propaganda exercise on the part of the Russian government.[8] Roberts, for example, argued

this was a most effective propaganda move. Unfortunately, neither we nor the Americans can safely indulge in this sort of political warfare since it is likely to confuse our own public and do more harm than good in softening up the enemy. The most important aspect of it is its encouragement given to appeasers of Germany and to the anti-Russian school in the USA.[9]

Eden agreed.[10] Clark Kerr was of the view that the composition of the committee ruled out the possibility that it was a nucleus 'round which a Communist regime might be formed after the war'.[11] There was a suggestion that the NKFD had been set up to annoy the British and American governments.[12]

By 1 August 1943, the Foreign Office seemed agreed that Molotov's statement that the NKFD was purely a propaganda device should be accepted. But it had raised far more issues than this. The Foreign Office believed that Allied policy on Germany after the war needed clarification to avoid a repetition of this sort of misunderstanding. Officially, it was accepted that the NKFD did 'not represent any political move on the part of the Soviet government towards a Communist Germany' because of the composition of its committee.[13] But there were also important doubts. First, there were real suspicions that the NKFD might be the first prelude to a separate peace between the USSR and the German army. 'We must be prepared', minuted Roberts, ' for German military opinion—which has always favoured the Bismarckian policy of good relations between Germany and Russia—to turn increasingly towards Russia with a view to a compromise peace.'[14] Secondly, British policy towards the exiles might prove to have been wrong:

We could persuade generals here to take similar steps if we had a similar propaganda policy and were prepared to recognise free German committees and I think we could secure their support a good deal more easily

[8] FO 371.34415 See above, p. 158.
[9] FO 371.34415 c 10020.
[10] FO 371.34415 c 8626.
[11] FO 371.34415 c 8488.
[12] FO 371.34415 c 8545.
[13] FO 371.34415 c 8616.
[14] FO 371.34415 c 1085.

than the Russians—we have however taken an entirely different line with the Prime Minister's speech of 21 September which attacked Prussian militarism as being as evil as the Nazi system.[15]

The Foreign Office seems to have had General von Thoma in mind, though other names were mentioned in this connection. There is no record of Rudolf Hess's name in this context.

There was wider public interest in Foreign Office policy as a result of the formation of the NKFD. On 24 July 1943 the *Times* carried a comment from the *New York Herald Tribune*:

if it does represent a political initiative by Russia in a field in which the US and Britain have been notably chary of committing themselves, one must gather from recent acts of western Allies that they are deliberately cultivating a political vacuum in Europe which is gradually to be filled under the benevolent eyes of the powers.[16]

On 30 July 1943 Eden was approached by the Bishop of Chichester, 'our good German bishop' as Strang called him.[17] He argued that the NKFD was based on Stalin's distinction between Germans and Nazis and offered the German people a definite hope for the future. But this was precisely its danger, Bell went on 'for the Communists controlled the NKFD though its programme is liberal and this was a far better way of Communist penetration than the way of force tried after the last war'. Bell feared that Russia and Communism would reign supreme in Germany unless the western Powers produced an alternative programme using their own Free Germans. Orme Sargent did believe that there was a good deal more to the NKFD than propaganda, but Strang argued simply that 'the spring of Bell's action is, as we have always suspected, the Communist bogey and he looks to an alliance with good Germans to meet that bogey. He and his like will lead us into a new war in half a generation if they are given their way.'[18] The Foreign Office, therefore, decided to pursue the issue along two lines. First, it once again asked MI5 to carry out detailed investigations into British exile politics and secondly, it determined to formulate a fixed policy which it could impress on its allies.

[15] FO 371.34415 c 1085.   [16] FO 371.34415 c 8626.
[17] FO 371.4415 c 8860. See R. C. D. Jasper, *G. Bell Bishop of Chichester* (London, 1967): Eden told Bell that it would not be in the national interest to ask him about Stalin's order of the day in public, pp. 273, 274.
[18] FO 371.34415 c 886.

There was little doubt that Wheeler-Bennett's reports on events in America caused anxiety.[19] He believed that Germans in exile would now be able to say, 'Why do you let Moscow have the monopoly of policy in regard to the future of Germany? You must give us a soft peace to prevent Germany from going Communist.'[20] He believed that any negotiations with representatives of the German army would be disastrous. Cadogan was interested in this theory. The argument was that if the German army were not ruthlessly destroyed, the war might as well have been won by Germany. He stated that the NKFD was being used by exiles in America to approach the State Department, the OSS, and the army with a 'refurbishment of the Bolshevik bogey' and that this policy had already met with success. 'The moderate school of thought in the State Department argues that the best way to fight Bolshevism is to deal with the people who are as anti-Russian as they are anti-Nazi.'[21] There seems little doubt that the Foreign Office were seriously disturbed by the prospect of having to negotiate in any way with Free German groups in Britain.

MI5's investigations appeared to support the view that Communists were successfully taking over the leadership of the German anti-Nazi exiles. And from this intelligence, it was of course possible to infer that the best way of countering Communist influence was to assist those groups who were not only anti-Nazi but anti-Communist as well.[22] A certain Thwaites, who was in charge of the German Section of the Ministry of Information argued strongly in favour of this notion. But the Foreign Office, despite the evidence before it, were for the most part unimpressed by the need for such a policy. Harrison noted:

MI5 exaggerates the danger that émigrés could fall under Communist influence. They—and Thwaites—suffer from so many delusions about the Austrian and German emigration that the Foreign Office will have to educate them.[23]

MI5 had investigated the FDB meeting at which Schiff threatened to resign unless Russian territorial aims were condemned.

[19] FO 371.34415 c 9879, 30 Aug. 1943.
[20] FO 371.34415 c 9879.
[21] FO 371.39415 c 10020, 2 Sept. 1943.
[22] FO 371.34416 c 13941.
[23] FO 371.39919 c 1370, 31 Jan. 1944.

Allen claimed that Vansittart's view of Schiff was being confirmed—he was a very 'black German'. Robert's view however
was that it was 'not unnatural that a German should take this
line'. The aims of the Soviets did cause real problems

although a good enough case can be made to satisfy British public opinion
in respect of East Prussia as the home of the Junkers, German militarism,
Danzig as Poland's port and Upper Silesia as being largely Polish by race
and industrially indispensible to Poland.[24]

Any German who now opposed the territorial rearrangement of Germany's eastern frontier could expect only to be
branded as a 'bad German'. Needless to say Lord Vansittart
was given ample scope to exercise his gifts. In April 1944 he
wrote to Strang about Friedrich Stampfer, .

I have been engaged in showing up some of the many mendacities of
Stampfer in the USA. I suppose you know all about this nasty piece of
work. He is in the full blown pan-German type masquerading as a Social
Democrat and is doing a good deal of damage ... I propose to call him
a typical German liar in public.

The erstwhile Chief Diplomatic Advisor to the British government went on, 'I don't think we should let these Germans get
away with this sort of thing in order to excuse the complete
inefficiency of their alleged underground movement in Germany.'[25] Stampfer, he wrongly alleged, was one of those Social
Democrats who held that the NKFD showed that 'Stalin's aim
is to spread a protectorate throughout Europe ... The Germans
will get a better deal from him than from the west.'[26]

On 6 January 1943 the Dutch Foreign Minister in exile in
London, M. van Kleffens had a meeting with Eden to protest
at the activities of *émigré* Germans. He had just returned from
the USA and he was fearful of the influence of men like
Treviranus and ' a certain former Prime Minister of Prussia
whose name he could not spell but thought was Grezenzinski
(*sic*) ... They had founded an organisation called Free Germans
Inc. a title which afforded M. van Kleffens considerable amuse-
ment.'[27] This incident led Harrison to request that the

whole question of relations with émigrès ought to be taken up at the
highest level in Washington ... The Americans think we are too lenient

[24] Ibid.   [25] FO 371.39919 c 5425.
[26] See, *Mit dem Gesicht nach Deutschland*, ed. E. Matthias (Düsseldorf, 1968),
pp. 618, 619.
[27] FO 371.34413 c 388.

with Strasser and we think that the Americans are inclined to flirt dangerously with German émigrés generally.[28]

There was undoubtedly some truth in the American view that
the Foreign Office showed more interest in exiles from the
centre and right wing of German politics than in those who
were members of the SPD. But it is worth pointing out that
some SPD members also accused the Foreign Office of a
marked pro-KPD bias, an allegation that gained significance
later on when it appeared that a number of Foreign Office officials had been Soviet agents. They would naturally have had
an interest in suppressing the SPD and supporting the German
communists.[29]

The Foreign Office was at this time considering British policy on Germany's future.[30] On 13 October 1943 Troutbeck
was appointed advisor on Germany and Con O'Neill was made
his assistant. These appointments were related to the fact that
the War Cabinet was shortly going to take decisions on the
political and territorial status of Germany after the cessation
of hostilities.[31] British plans towards Germany require some
explanation, therefore, not least because the severity of Allied
proposals shows why there was no room in Allied counsels for
exile plans. On 28 October 1943 Eden sent his colleagues in
the War Cabinet a memorandum 'on the future of Germany'.[32]
Eden was shortly to leave for Moscow and despite Churchill's
order that 'nothing about Germany should be fixed', Eden
needed a policy to take with him. He saw three alternative
plans as feasible. The first was that there should be a unitary
Germany with a central government. The second was a decentralized administration on a federal basis, and the third was
the splitting up of Germany into separate states which involved the British 'doing all they could to encourage separatist
and particularist movements in Germany'. The War Cabinet's
view was that Germany should become a 'number of separate
states and separatist movements were to be encouraged', but
that 'it was impossible to say whether it was possible or feasible to force it on the Germans'. Whether or not there was to

[28] FO 371.34413 c 319, Feb. 1943.
[29] See above, p. 159.
[30] FO 371.34461 c 12101, Oct. 1943.
[31] Ibid.
[32] FO 371.34461 c 12635 (marked 'Most Secret'), 28 Oct. 1943.

be a central government, Germany was to be administered for as long as two years by an inter-Allied body. Germany was further to revert to her pre-*Anschluss* borders in the south, be totally disarmed, lose East Prussia, Danzig, and considerable areas of Silesia with a transfer of the German population and her industry was to be subjected to international control.

In September 1943 Eden had already informed Molotov that 'the UK particularly wished to see the separation of Prussia from the rest of Germany'.[33] In Moscow, however, Eden found it hard to get full agreement from Molotov for this, or indeed to gain any firm statement of Russian views on Germany's future. When asked whether he agreed that Germany should be dismembered Molotov stated simply that the Soviet government supported any move calculated to make Germany harmless as an aggressor but that there should be no public statement about this because 'the Germans are still on Russian soil and it will make them fight harder'.[34]

American policy was considerably less harsh. The US Secretary of State favoured a 'broadly based democracy, operating under a Bill of Rights to safeguard the civil and political liberties of the individual'. Among conditions for the success of a new democratic experiment in Germany would be a

tolerable standard of living, early steps to restore freedom of speech, religion, press, freedom to organise political parties other than Nazi or fascist ones, and where conditions permitted, to make preparations for free elections.[35]

What these views showed was important. It was clear that the Americans were far more liberal and progressive, not to say realistic, in their plans for Germany after Hitler.[36] The British, on the other hand, had very severe measures in mind, ones which no political exile who hoped to gain power through democratic means in post-war Germany could possibly support. Finally, the Russian policy was far more mysterious than either of the other two. Molotov was determined to keep his options open. He refused to commit the USSR to the British

[33] FO 371.34461 c 12645.
[34] FO 371.34461 c 13589.
[35] Ibid.
[36] See Woodward, op. cit., (1976), pp. 76, 77. It appears that Eden was committing the UK to structural changes in Germany which Hull did not want.

course and he showed a characteristic concern with the propaganda dimensions of Allied policies. At the same time, however, the Russian refusal to state exactly what the USSR hoped to see, could be seen as highly alarming. It was feasible that the USSR was toying with the idea of a separate peace with anti-Nazi Germans or at any rate that it was prepared to encourage the Germans to think this was possible. The NKFD was obviously a platform in the policy and it was one the British were not disposed to view with equanimity.

In November 1943 a document was drawn up by the Foreign Office which set out British policy towards the NKFD. It was printed and circulated on 31 December 1943. The British had drawn a distinction between Germans and Austrians, it was recalled, but this had not produced any tangible results simply because the 'overwhelming proportion' of exiled Austrians were Jews who had no wish to return and were 'totally unrepresentative of present-day opinion in Austria'. There was no change in the fundamental policy 'that it was not desirable to promote the organisation of a Free German movement'. The announcement of the formation of a Free German movement in Moscow in July had 'provoked endless discussion and in Great Britain German Communists launched a free movement to collaborate with the Moscow organisation'. But it had been regarded with suspicion by most of the other groups here who 'have failed to respond to it'. Thus, the report concluded, HMG did not propose to give any encouragement to any free German group or individual. They were 'unlikely to gain influence to cause HMG's policy to change', and it was worth recalling that the Soviets had said their committee only served propaganda purposes.

HMG's attitude was thus completely plain. There would be no recognition of any exile group. British policy with regard to Germany's future was not to be formulated without any reference to political movements that existed in exile. Even if there could be seen to be a political opposition to Hitler inside Germany, HMG did not believe that it was in the British interest to support any political movement which might lead to anything less than a radical reorganization of German politics and geography.

These views found expression in one further document of

interest, the memorandum by the Foreign Office for Eden to take with him to America for a Foreign Minister's meeting in Washington. Dated 13 February 1944, it pointed out that *émigrés* like Treviranus, Rauschning, Grzesinski, and Brüning were able to influence public opinion in America. Their aim was to secure 'tender treatment for Germany' by the Allies. This was extremely dangerous because their European allies, especially the Poles and the Dutch, regarded such a policy with grave suspicion. Any concessions to *émigrés* in the USA would produce immediate repercussions in Britain. In November 1942 the British Embassy in Washington had been asked to raise this matter informally and, although the State Department had said that 'it was unlikely that any concessions would be made to exiles', HMG were still anxious.

The British reasoned that none of the *émigrés* had any following in Germany or was likely to give substantial assistance to the Allied cause. Any concessions were likely to lead to the demand for the recognition of a Free German movement. HMG had been very careful to abstain from giving any encouragement to *émigré* Germans and furthermore 'the anxiety expressed by the Soviet authorities in connexion with Rudolf Hess is only one illustration of the sensitiveness of our allies in this matter'.[37] Any concessions would simply encourage these groups to redouble their efforts, especially when Germany's defeat became imminent. The Foreign Office was clearly anxious lest there be any lobbying of the United Nations over peace terms. The final argument against any official contact with the exiles illustrates yet again the central problem of the existence of an organized opposition to Hitler in exile—'there can be no positive advantage in (talks) since no émigré is looked to for leadership and any appearance of Allied support might produce a result the reverse of that desired'.

It is certainly possible that better information about the past history of the SPD in exile might have tempered British attitudes. MI5 gave a good account of the present state of the party but it could not convey the significance which the party had possessed and which it was determined to possess again in the future. Finally, in this context, there is the chance that information favourable to the SPD may have been suppressed

[37] FO 371.34413 c 3668.

by officials in the Foreign Office who wished to see the Communists prosper.[38]

The Foreign Office, then, saw absolutely no reason to alter its policy towards German political exiles in general and towards the SPD in particular. The fact that the Soviet government was following a different course did not appear to cause undue anxiety about the wisdom of the British line. Stalin was thought either to be making rather crude propaganda which Britain could not have done even if it had wanted to because of British public opinion, or else it was seen as a subtle hint to precipitate the setting up of a second front in Europe. There is no evidence to suggest that the Foreign Office as a whole believed the Soviet government was creating a puppet government in exile which it would install in office after the cessation of hostilities, even though this was perhaps the most likely explanation for the recognition of the NKFD.[39]

As far as the Foreign Office was concerned a harsh policy towards German political exiles seemed quite proper and, as the war drew to a close, its policy if anything became even more harsh. When one particular group organized an exhibition entitled '10 years of Hitler-fascism' Eden himself wrote that he wanted 'something done about these people, exposing them if not suppressing them altogether'.[40] The only reason that anyone appeared to show any interest in the activities of the SPD (many intercepted letters and reports of Social Democratic meetings were kept on file) was the MI5 view that 'they could fall under Communist influence'. In November 1943 MI5 actually asked the Foreign Office to agree to the licensing

[38] See below, p. 203. See P. Seale and M. McConville, *Philby, the Long Road to Moscow* (London, 1978), p. 218: 'There was the suppression on two occasions late in 1942 and again in the spring of 1944 of evidence suggesting a Conservative opposition wished to make contact with the British in the struggle against Hitler. On the first occasion, Philby refused without explanation to allow the circulation of a paper prepared by Trevor-Roper's RIS . . . on the second he dismissed a warning of the July 1944 plot . . . given by Otto John.' The implication is that this was done on Russian orders. See below, p. 201; Lord Sherfield, in an interview on 8 May 1978 argued this theory was complete nonsense. However R. V. Jones in *Most Secret War* (London, 1978) points out that Philby did have a key position, pp. 254, 255.

[39] See Woodward, op. cit. (1976), pp. 22, 209. In 1944 Sir O. Sargent began to express a fear that the USSR would try to install a puppet government in Germany.

[40] FO 371.34414 c 5329, 12 May 1943.

of a non-Communist newspaper intended for German workers
to be called *die Arbeit* since 'it was not in the British interest
to let the Communists and Russians control' all German exile
politics.[41] W. D. Allen, a junior official, suggested the Foreign
Office should support MI5's suggestion that 'it is not in our
interests to deny facilities to the Social Democratic opponents
of Communism'.

But Harrison strongly disagreed: 'these émigrés are really
very unimportant', and he claimed that MI5 was 'exaggerating
the danger' of Communist influence. On Allen's insistence,
however, a licence was granted but it was to be revoked if it
'started to propagate obnoxious pan-German and soft-peace
material'. It is certainly true that the Foreign Office appears
to have seriously underestimated the control that the Communist party be received to hear what they have to say', and
1943. It is also perhaps a little surprising to find in the record
that on 1 June 1944 a leading German Communist, Koenen,
had his request for a meeting with Eden carefully considered
by the Foreign Office.[42]

Harrison urged that someone received Koenen and a colleague and he suggested Lord Rea was the right person. 'It is
quite appropriate that representatives of the German Communist party be received to heat what they have to say', and
Lord Rea arranged to see them on 8 June. There is, alas, no
record of that meeting but the SPD was not similarly treated.
If it was appropriate for the KPD to see a high Foreign Office
official, it is not illogical to suppose that it would also have
been appropriate for the SPD. The fact that it did not come
off may, then, be attributed to a certain anti-SPD tendency.

If the Foreign Office had known how significant the SPD
was likely to be (and indeed became) in post-war Germany and
that it would be a useful bulwark against the westward spread
of Communist influence, the SPD might have received a more
impressive part to play in British policy.[43]

Public opinion, of course, may have played a more important role. It is significant that the Foreign Office read with

[41] FO 371.34416 c 13941, 25 Nov. 1943.

[42] FO 371.39062 c 7331.

[43] See Woodward, op. cit. (1976), p. 22. Foreign Office memorandum of
8 March 1943 spoke of 'the need to convince the German people that their best
interests lay in the West'.

interest a report produced by the British Institute of Public Opinion on 7 October 1943 on the British public's feelings towards Germany. This appeared to support Foreign Office hostility towards any notion of 'another' Germany. In reply to the question 'what are your feelings towards the German people?' forty-five per cent said they were 'bitterness, hatred and anger', twenty per cent 'the Germans are getting what they deserve' or 'dislike of the Germans', whereas only fifteen per cent said they had feelings of either 'friendliness or pity' towards the Germans.[44]

The Generals' plot against Hitler on 20 July 1944 seems to have changed British thinking towards Germany very little, even though, for better or for worse, it proved that not all Germans fanatically supported Hitler. The main conclusion that Harrison seemed to draw was that the failure of the plot confirmed the correctness of the British firm line towards Germany. Now more than ever, he suggested, the German people would follow Hitler. Eden thought this 'very interesting' and he agreed with the notion 'that now the Nazi regime will be riveted even more firmly on the neck of the German people . . . Himmler's control will be even more complete'.[45] Ironically, perhaps, Eden's actual argument suggests that he was beginning to take a more sensitive view of the Nazis' hold on the German people, although, of course, his conclusions meant this was of no consequence. Even if it was true that the German people supported Hitler only because they feared him and his regime, Eden still saw the entire destruction of Germany as the only means to break that hold.

As the war moved towards its end, it seems that even the merest consideration of utilizing or even encouraging German political exiles, especially members of the SPD in any corporate way, disappeared. Certain Social Democratic leaders continued to work as individuals for Hitler's defeat—Eichler with the BBC, Fritz Heine with SO1, Ollenhauer with OSS—but there was absolutely no attempt at any systematic co-operation. This was not, it should be stressed yet again, a matter of simple

[44] FO 371.34461 c 12764, 16 Nov. 1943.

[45] Wheeler-Bennett wrote 'it is good that the plot failed otherwise the generals might have stayed in power. It is possible that Himmler may now try to overthrow Hitler and negotiate.' FO.371.39062 c 9773, 21 July 1944.

logic. There was every reason why the Foreign Office ought to have considered co-operation with the SPD in greater detail. After all, the general plans for Germany that the Foreign Office was drawing up seem to take the continued existence of organized labour movements not only as a given political fact but as a desirable one too. A pre-surrender guide for Germany, for example, produced on 3 February 1945 contained the general directive that free trade union movements were to be encouraged.[46]

The apparent poverty of Foreign Office policy towards Germany can be seen in three crucial areas. First, there was serious lack of information and understanding. Second, no attempt was made to build up the SPD's strength to help it counter the Communists. And, third, because of these two factors British influence on domestic German developments was very limited. When we recall that the origins of the Second World War lay firmly in domestic German politics, the Foreign Office's position is less than understandable.

By early 1945 those reports being received in Whitehall all pointed to the danger that a weak SPD would almost certainly succumb to a vigorous KPD. It is worth noting that the Soviet hold over eastern Germany was made possible only because the KPD had been able to dominate the SPD in that area. Firmer British support for the SPD might have strengthened its resistance to Communism—and half Germany might have been spared Communist dictatorship.

It is important to realize that ignorance alone does not explain the Foreign Office's stance. Two other factors played a part. The first was apparent British anxiety not to alienate the Soviet Union throughout this period and into 1946. The SPD was fundamentally a friend of the West and specifically akin to the British Labour government. The USSR was therefore bound to suspect that a supportive attitude towards the SPD might be merely a hostile move against itself. William Strang expounded these points in a memorandum for Ernest Bevin at the end of 1945.[47] He reminded him that at Potsdam it had been agreed to hold *Kreis* elections in May and June 1946 as part of 'our deliberate policy to encourage party activity

---

[46] FO 371.46827 c 403.
[47] FO 371.46910 c 101, 28 December 1945.

on a wider basis'.

At the same time, however, anything too specifically contentious was to be avoided. For example, the British 'could not allow issues like Germany's eastern frontier to be discussed in public'. Finally Strang pointed out that British policy towards the SPD had become a critical matter:

> We should probably like to see a strong SPD but its key position in the political spectrum would be likely to make it a serious bone of contention between the Allies and it may be as well that the present equivocal situation should continue (even though) the SPD represents the centre of gravity in the British zone and also probably in Berlin and its attitude is most nearly in accord with the general feelings in western Europe.

Strang concluded that if the SPD supported amalgamation with the KPD in Berlin its position in the West would become 'intolerable' but that if it seemed 'Western based', the Soviets would refuse to give it any power. The SPD's position was 'very awkward' and so it was best for the British to do nothing.

Strang's logic was certainly sound but his policy was clearly defeatist. If the SPD gained firm British support, the Russians would ignore it, he argued. But if it remained weak and susceptible to Communist domination, the Russians would take account of it. Thus, either way, Russian interests would be served. The third alternative: to challenge Russian plans for domestic German politics, by aiding the SPD so that it could withstand Communist pressure, was not considered.

Ignorance and an unwillingness to challenge the Soviets were two factors which may account for British policy at this time. One other one must be borne in mind: treachery. There is some evidence that within the Foreign Office and the British secret services there were a few influential though not very senior officials who wished Communism to triumph in postwar Europe.[48] Their efforts were helped by the fact that the distinction between British interests and Soviet interests was not always a clear one in the minds of British policy-makers. Whilst it was clearly vital for the Third Reich to be overthrown if Britain were to survive, post-war Communist penetration in Europe was hardly a desirable British war aim. At the same

[48] See Seale and McConville, op. cit.; A. Boyle *The Climate of Treason* (2nd ed., London, 1980); H. R. Trevor-Roper, *The Philby Affair* (London, 1968); B. Page *et al.*, *Philby* (London, 1977) and H. A. R. Philby *My Silent War* (London, 1968).

time, of course, not everything that aided the Soviet Union in its war effort was disadvantageous to the British.

Against this background, it seems plausible to suspect that Philby and his comrades were able to exert some influence on British policy. Their strategy, both as intelligence gatherers and as advisors, was to do nothing which might place Communists at a disadvantage in the post-war world and to sow confusion amongst their colleagues about the aims of Communists in relation to those of Social Democrats and other anti-Nazi groups. In order to advance the fortunes of the Communists, then, they were depicted as wanting a Europe not incompatible with British interests and sufficiently strong to lead all the forces on the left of politics. Men like Philby clearly understood that the policy of the Foreign Office towards anti-Nazis was a considerable advantage to the Communists. At the same time, however, it would be unwise to exaggerate the role played by Communist agents or to suggest that every British policy was undermined by them.[49]

As it was, a weak and demoralized SPD which had had to fight not only for the 'other Germany' but also for itself against the British Foreign Office (and the British Labour movement) was hardly the most robust opponent of Russian-tended German Communism. Indeed, independent information 'from a reliable source', possibly the American Secret Service, reached the Foreign Office on 9 June 1945 and made precisely this point. It argued that the SPD would 'either yield to popular front pressure'—that is coalesce with the German Communists —'or disintegrate'.[50] The reason for this was, in the words of the report, 'the fact that the British support that the SPD counted on has not materialised'.[51]

The SPD had, it was argued, for years been attempting to fight off the Communists' threat to its existence. Vogel and Ollenhauer in London had been attempting to maintain an independent line from the KPD whilst at the same time refusing openly to alienate the Foreign Office by too strong criticism of British plans for Germany. The behaviour of the

[49] Trevor-Roper, op. cit.
[50] FO 371.46910 c 6122, 10 Sept. 1945.
[51] This was also Kurt Schumacher's view; cf. A. Kaden *Einheit oder Freiheit* (Hanover, 1964), p. 20.

Stampfer group in the USA was cited as evidence of what the
SPD in London might have been like had the SPD leaders in
London not shown such reserve. The report noted that they
now appeared to fear 'being left high and dry' and it warned
the Foreign Office of KPD intentions. Communists, it was
said, were 'rushing to join the Anglo-American forces as inter-
preters' and the KPD in London had formulated precise policies to be executed on its return. These included support for
a government in the 'separate Russian zone if a united government for the Allied zones is decided upon'.[52] The National
Committee for Free Germany (NKFD) was to become 'the
national government and the people's army' and as soon as
USSR rule over its German zone 'was assured there will be a
great propaganda drive for national unity'.

The facts proved to be remarkably similar to the predictions
in this report. Ever since Soviet policy towards Germany had
taken a different direction from British policy, such a development was a clear possibility: this, indeed, has been the argument of the preceding chapters.[53] The British attitude towards
the 'other Germany' had been unwise. It had, in part, led to a
bankrupt attitude towards post-war German politics, an attitude which allowed the Soviet authorities to seize and hold
the initiative.

The fact remains that it was precisely when a more encouraging policy towards the SPD in exile ought to have been producing a dividend for British interests, that the British were
least able to show any mastery of internal Germany developments. When on 25 July 1945 Hans Vogel wrote to the Foreign
Office to allow the freedom of association in Germany as an
antidote to the permission granted to 'anti-Fascist parties in
the Soviet zone' all the Foreign Office could think of saying
was, apparently, summed up in Con O'Neill's note that 'Vogel
has always claimed to be the head of the SPD. In view of this,
it is rather odd that he is not pressing to return to that country
and his title is going by default.'[54] This was an utterly ignorant

[52] Cf. Woodward, op. cit. (1976), p. 209.
[53] When the British Embassy in Moscow noted the similarities between KPD's
manifesto and that of the NKFD, the FO view (in June 1945) was that it was
'simply a propaganda device' in order to seem 'bourgeois'. Molotov claimed he
knew nothing of these matters; 46910 c 3443.
[54] FO 371.4691 c 4163.

comment but this is really not the point—Vogel (who was in any case to die within two months) was not trying to further his own personal ambitions but trying to make the Foreign Office understand that there was a real danger that Communist influence in Germany might be decisive. Vogel argued that it was 'an urgent necessity ... to build a new Socialist Democracy in Germany and lead the German people back to the community of peace-loving nations'. There should be free political activity for all democratic German groups, he went on to insist, and he attempted to show how the entire political structure of the SPD was built on its belief in democracy. This was why the party needed to be 'rebuilt' so that it hoped that soon 'the conditions will be created for holding a conference of the whole party to decide on the policy and the leadership of the resurrected social democratic party'.

Con O'Neill was therefore not only badly misinformed (since Vogel did wish to return), but failed to appreciate the most important matter namely the re-creation of a powerful democratic grouping in Germany in order to counteract Communist plans. The Foreign Office certainly knew from first hand information of Communist overtures towards German Social Democrats who had not been in London. On 19 June the SPD and KPD met in Berlin and 'firmly intended to co-operate to liquidate Nazism, to create a bloc political party, to have joint political meetings on behalf of all working people and to discuss ideological issues'.[55] The Foreign Office received this announcement in the form of a flysheet from Berlin. Similarly the Foreign Office gained notification of the formation of a united anti-Fascist front in Berlin on 14 July.[56] There is no record of any British reactions to these events and no clear indication that the Foreign Office even saw the effect of these events on British interests and future policy.

Much the same point can be made about William Strang's report on the Wennigsen meeting of the SPD which the Foreign Office received on 27 October. This meeting, held 5–7 October 1945, was the most important gathering of Social Democrats in Germany since 1933.[57] Yet he gave no analysis of what he

---

[55] FO 371.46910 c 4167.
[56] *Ibid*; cf. A Kaden, op. cit., for a detailed discussion of this period.
[57] See below, p. 240 ff.

had seen and explored none of the implications of this conference. He noted that Schumacher had challenged the authority of Grotewohl supported by the 'leading delegate from London, Ollenhauer, who made an impressive appeal for unity but emphasised that the other non-Nazi parties were not necessarily the friends of the SPD'. Schumacher, however, was simply dismissed as 'dishonest' because he refused to explain that the difficulties with the British military government had been 'his fault'.

It was, it seems, only in November 1945 that Schumacher's importance began to be appreciated.[58] At a meeting of the SPD in Cologne to which he spoke on 12 November the Foreign Office noted that this 'was the first speech by a German political leader on foreign affairs'. It was significant that he did not accept the eastern frontiers and appealed for 'German survival which inside Germany will probably win unanimous support'. O'Neill wrote that he hoped the SPD would not accept Schumacher's views as its policy although he accepted that Britain could not 'have things both ways by encouraging the creation of German parties and then expect them to say nothing we dislike'. William Strang added his impressions—that Schumacher was now the acknowledged leader in the British zone and that he wanted the Germans to tackle reconstruction without British help. He would not accept that all Germans were guilty for the crimes of Nazism—for there were Germans in concentration camps at a time when others were still negotiating with Hitler. Finally, he related that Schumacher had said that 'guilt rested solely with those who missed the significance of class warfare, for the left discovered the theory and the right put it into practice'.[59]

In Berlin too it seems that by 20 November the British Foreign Office believed it had been too tardy in understanding the significance of political events there. On 11 November Hynd, the Government Minister responsible for Germany's British zone, went to Berlin in order to have discussions with political leaders. Strang said he had had 'an excellent effect in stimulating political life there *when hitherto the Soviet*

---

[58] FO 371.46910 c 8825; cf. Kaden, op. cit., p. 72.
[59] For Schumacher's view of class conflict cf. *inter alia* H. K. Schellinger, *The SPD in the Bonn Republic* (Den Haag, 1968).

government alone has appeared to take an interest'.[60] This was especially true in the case of the SPD 'who are under constant pressure to draw closer to the KPD and who should be encouraged to maintain their stand for an individual programme'. Strang pointed out that the SPD would serve the British interest well since it 'was not only numerically stronger than the KPD but the Berlin party is progressive and enjoys the inestimable advantage so far as the election is concerned of not being identified with the USSR'.[61] There was, he stated, no reason why the British authorities ought to 'abstain from encouraging the SPD's independence'. A minute on this report stated the time had now come for Britain to 'stop sitting on the fence and involve itself more actively to stop the SPD swinging to the left' but Con O'Neill disagreed—non-intervention in Austria had been 'very satisfactory from our point of view'.

It is plain, therefore, that once the importance of events concerning the SPD began to strike the Foreign Office, its analytical and policy-formulating concerns began to manifest themselves. But there is evidence, that this came rather late and that there was too little of it. Attlee (who had of course been British Prime Minister ever since the summer) claimed that the only way 'he could get a really clear picture of what is happening in Germany' was by reading the letters his own nephew had sent him (his nephew was stationed in Germany) and he passed at least one of them on to Hynd.[62] Hynd's apparent lack of interest in understanding German politics was also noted. In March 1946 Attlee received a letter from Geoffrey de Freitas complaining about Hynd's understanding of German politics. Hynd, it was said, 'is hardly ever in Germany and all members of the House knew this'.[63] The British control commission in Germany, Attlee was told, was 'dangerously short of staff and everyone is being urged to join it', and that 'Germans will have to be given a good many responsibilities which they were not intended to have'.

The British Prime Minister was urged to 'send to Germany a man of wisdom, courage and political experience to take full

---

[60] 46910 c 8989 (my italics). It was rather late to make this point so innocently.
[61] Cf. Kaden, op. cit., p. 287.
[62] Attlee papers Box 2.   [63] Attlee papers Box 7.

charge of the administration'. The present system was an 'insoluble enigma to the Germans. Sometimes it looks as if Monty is in charge, sometimes General Templar the driving force in the commission. And sometimes, though rarely, Mr. Hynd'.[64]

One of the first attempts by the Foreign Office to make some kind of policy statement on the basis of German political developments came with Noel Annan's investigation into German political parties.[65] Marked 'confidential', it explored the state of German politics in December 1945. Parties were not yet 'flourishing' he alleged and 'party politics are unrealistic because of the occupation'. The basic struggle was still one for existence 'yet in spite of this it is surprising how quickly the old parties have revived'. This was perhaps more surprising to Annan than it needed to be simply because a more balanced assessment of exile politics would have suggested that this was a possibility. 'The SPD and KPD have displayed remarkable vitality', Annan went on to state, and he added that 'the KPD is the most powerful party in the eastern zone and least powerful in the western Germany. All the *antifa* organisations have been crushed by the military authorities.'[66] Although the SPD had emerged as 'potentially the most powerful party in Germany' all was not well for it. There was, Annan alleged, a conflict between its left and its right and in the British and American zone party 'elders like Severing, Schoenefelder

[64] Attlee papers Box 7. There were, of course, many criticisms of British policy towards post-war Germany apart from this. Josiah Wedgwood, for example, drew Attlee's attention in June 1946 to the need for food for Germany to be released from government warehouses as soon as possible. 'Why should food parcels not be sent to the Germans?' he asked, for he and his workers got more than enough from the US; 'if the answer is that charity must stay at home, we know that to be the most immoral and rottenest answer which was used to keep out refugees in the 1930s who subsequently were murdered at Auschwitz'. Attlee replied rather as Wedgwood suspected he might, saying that it was 'inopportune to send food parcels out of the country when we are having to ration bread' and his secretary minuted 'the PM is anxious not to enter into any detailed controversy with him or sign anything which would help him in his campaign'.

[65] FO 371.46910 c 10128.

[66] See *Arbeiterinitative 1945*, ed. by L. Niethammer *et al.*, (Wuppertal, 1976). On 12 December 1945 Willi Eichler asked permission to be allowed to return to Germany to edit an SPD paper in Cologne. The FO noted that he had been in Britain since 1940 and 'made himself useful to the British authorities in various ways. He is the leader of a small unofficial group of Social Democrats in this country and if we allow him back we shall have to allow some of the official ones as well.' This meant Ollenhauer, who was allowed to return early in 1946. He had of course been given permission to attend the Wennigsen meeting.

and Hennerath had taken over. Severing's past is shady and he is utterly discredited with the rank and file and a *persona non grata* with the British authorities.' All Schumacher was doing was waiting for the right time to challenge Grotewohl. Much of the SPD doctrine Annan concluded was 'orthodox— land reform, nationalisation of large industries and so on. But Schumacher also wants to win over the middle classes.'

When in December 1945 Maurice Edelman MP wanted to know what British policy in Germany was, the Foreign Office replied 'it is to foster the formation of parties on the *Kreis* level because this is the best means of developing a sense of the true meaning of democracy, by building from the bottom up'.[67] A junior official wondered whether it was right to offer such a meaningless reply which in reality was simply 'a parrot cry—it sounds good but means little'. In some senses this is the right epithet for all of British policy towards German political exiles throughout the Second World War.

Even the view that British support for the SPD in exile would have made it seem to many Germans a party of traitors cannot be considered wholly convincing. Indeed even if some Germans had considered that some of the party leaders had behaved treacherously, the party as a whole might not have been harmed. But this was never the serious basis of any Foreign Office policy: it was not being 'cruel to be kind'. Soviet ambitions, were ably served by Soviet policy, and this showed how effective another course of action might have been. Foreign Office reports for 1945 express the consequences of the poverty of British policy in this area.

It seems fair to consider whether the subsequent division of Germany could have been avoided had British policy towards the SPD been more generous. The creation of two German States was a source of grave concern both to those who wished to see a free and united German nation and to those who longed for peace. That Germany was divided was, thus, in a variety of ways, to be greatly regretted. If a strong SPD, sup- ported by the British, had been given the same opportunities that the Soviet Union gave to German Communists, the div- ision of Germany would have been far harder, though not im- possible, to uphold. It is perfectly true that this division was

[67] FO 371.56910 c 9230.

primarily based on the military power of the Red Army on the one hand and the U.S. and British forces on the other. But occupation itself depended on a political substructure, so that even if Communist domination might not have been successfully opposed by a strong SPD, it would undoubtedly have made it more difficult to establish. It is just possible that a challenge, early on, might have encouraged the Soviet Union to re-think its German policy. A stronger SPD might have been a small but significant part of such a challenge.

British attitudes towards German political exiles were subordinated to the desire to destroy the Third Reich. This was in many ways quite justified and alternatives would have produced a variety of problems. But, in the final analysis, because the German people were not going to be destroyed even if their government, their houses, and their jobs were, it was a limited policy and one less realistic than its allegedly more idealistic alternatives.

X

The SPD and the National Committee of Free Germany

ALTHOUGH the plans and proposals presented by the SPD leaders in London demonstrated that German Social Democracy had not been destroyed, there seems little doubt that by the middle of 1943 the situation of the SPD in exile was rapidly deteriorating. A balance sheet drawn up at this time would have shown that on the credit side it had been able to survive in sheer physical terms in that its leadership was still alive, and in that it had produced a number of policies. On the debit side, however, it had failed disastrously in its attempt to influence the formulation of British policy towards Germany either through the Labour party or the Foreign Office, and it had been able to undertake no real part in the fight against the Nazis. In 1942 the credibility of the exiled party had been seriously undermined by the resignations of Geyer and others, and the fact that the Labour party now refused to fund Ollenhauer and Heine meant that by 1943 the opportunities for any serious political activity appeared practically non-existent.

Unless the SPD could once again demonstrate that there was proper work for it to do and that its political credibility could be restored, it was bound to become a mere ghost, a pathetic shadow of a once great political organization, existing only in the minds of a few unfortunate and argumentative German refugees. But, rather surprisingly, the SPD did not go under. It was able to show that it had a function to fulfil in exile and it was able to prove that there was a good political reason for its continued existence.

It was not without irony that the SPD was saved from premature burial by one of its would-be undertakers, the KPD. German Communists had already on a number of occasions attempted to destroy German Social Democracy. Although they had been unsuccessful in 1919, the efforts made by the KPD in the period after 1929 to undermine the stability of the Weimar Republic—when it had even co-operated with the Nazi party—finally bore fruit in 1933. Whatever significance is attached to the ideological contest between the SPD and the KPD, at the root of their struggle lay their con-flicting claims to the allegiance of the German working class. The control of the minds of this social group appeared to offer the chance of controlling the German state.

When, therefore, on 12–13 July 1943 a National Committee of Free Germans or NKFD was formed in Moscow, the SPD leadership in London assumed that it was yet another attempt by the KPD to gain control of the German working class and eventually of the post-Hitler state.[1] This suspicion was fully confirmed by the fact that in June 1943 the representatives of the KPD in London had approached Vogel and Ollenhauer with a view to setting up a Free German movement—or FDB —in England which would seek to heal the rift between the two aspiring leaders of the German working class.[2]

The formation of the NKFD raised three important issues. First, it appears that Stalin's propaganda line, that a distinction between Nazis and Germans could and ought to be drawn, was beginning to produce practical political results. Secondly, what appeared to be a German government in exile encompassing groups ranging from the KPD to the aristocratic right, was in fact dominated wholly by Stalin and the Communists. Thirdly, it seemed possible that by virtue of the power of the USSR and the Red Army such a shadow government might not only stand an excellent chance of gaining political power in post-war Germany but it might also be able to incorporate and destroy the SPD. There was no doubt in the minds of the SPD leaders in London that the NKFD represented the gravest threat to the future of the SPD since the German army had advanced on Paris in 1940. They knew their stock was ex-tremely low with the British authorities and they therefore deduced that any Free German committee would command political importance after Hitler had been defeated.

[1] See above, p. 18.
[2] SPD *Mappe* 150, 15 Dec. 1943, letter from Vogel to Rosenfeld.

This was the real nadir of exile. External political credibility had been dissipated; internal unity had been badly mauled; and now, in 1943, the one prospect that remained, the future, seemed to be taken from them as well. The leaders of the SPD in exile faced yet another test of their political courage. To hold out against the KPD, to resist its warm red embrace, would most probably prove the final nail in the SPD's coffin. Not to hold out seemed but another way of hammering in the same nail. What they could not know was that the policy they eventually decided to pursue was to contribute as much if not more than any other single exile policy to the continued existence of the SPD. By refusing to co-operate with the KPD in the formation of the Free German committee in London, the SPD was able to maintain a political independence which stood it in very good stead once Stalin's ambitions towards the German people became plain for all to see.

It is worth noting that the British Labour party was faced with a very similar dilemma at this time. On 18 December 1942 Harry Pollitt, for the Communist party of Great Britain, officially asked the NEC of the Labour party to allow it to affiliate with the Labour party.[3] On 18 February 1943 Middleton wrote to refuse this request largely because its affiliation to the Comintern would mean that the CPGB would not be able to accept the decisions of Annual Conference. On 28 May 1943, following the dissolution of the Comintern, the NEC had a special meeting to see whether the Communists might now be eligible but decided that they were not.

Röder has suggested that the formation of the NKFD was the culmination of the popular front tactic initiated by the Comintern in 1935.[4] Although it is not easy to provide satisfactory interpretations of any Soviet policy during this period, this view, despite being an oversimplification, is in essence not incorrect, notwithstanding the fact that the Comintern had been dissolved in May 1943 before the NKFD had been formed. The point was that meaningful negotiations between Social Democrats and Communists had always foundered on the Social Democrats' refusal to accept that Communists could act independently of Moscow. The KPD, for example, was

[3] The Annual Report of the Labour Party Conference for 1943, pp. 9-19.
[4] Röder, op. cit. (1969*a*), p. 198.

seen as the simple agent of Soviet power and the Comintern was nothing other than a Stalinist instrument of imposing policies on member Communist parties.[5] Thus the dissolution of the Comintern was seen by the USSR as a necessary precondition for the formation of popular front pacts and broader coalitions. Communist parties would no longer be seen as the means of spreading Bolshevism throughout the world but as parties with national identities.

The formation of the NKFD was correctly seen by the SPD in London as a new departure in Soviet foreign policy. It was now no longer the case that Communists wished to coalesce with Social Democrats alone but that, by virtue of the power of the Red Army, Communists felt strong enough to dominate a wider political coalition.[6] The aim of the popular front tactic had been to heal the rift in the international Labour movement which had allegedly caused its defeat by Fascism, to produce 'unity of action by the working class so that the proletariat can grow strong in its struggle against the bourgeoisie'.[7] The NKFD on the other hand, was more grandiose in conception and it was later to prove a solid foundation for the Soviet control of the East German state, the German Democratic Republic. Participation in such a movement would be extremely hazardous for the SPD. Yet non-participation might also prove very damaging. After all, it was difficult for the SPD to claim that it represented the 'other', democratic Germany, and then refuse to have anything to do with a freely constituted body which was composed of all the remaining anti-Nazi groups. Isolation from a progressive and effective political body would make the SPD appear even weaker than it had already become. If it wished to be taken seriously as an ally in the fight against Hitler, it could hardly refuse to join such an impressive anti-Nazi coalition.

Ever since 1935 SPD leaders had faced the embarrassment of overtures from German Communists with a view to some sort of pact. But the SPD had always been able to resist these advances arguing that Communist ideas were as reactionary and

[5] See above, p. 101.
[6] See Matthias, op cit. (1968), p. 241 ff.
[7] G. Dimitroff, *The United Front* (London, 1938), pp. 15, 32; Duhnke, op. cit., pp. 163–82.

totalitarian as National Socialist ones.[8] The SPD had always pointed out that the KPD had helped Hitler destroy the Weimar Republic because it believed that Social Democracy was the real enemy of the working class. The Hitler–Stalin pact of August 1939 brought considerable relief to the SPD. Negotiations were at once ruled out and ideological points could be scored in great number. It confirmed the SPD's thesis that extreme left and right were, in fact, natural comrades.

But Hitler's attack on the USSR reopened the old arguments, and the heroism and the success of the Red Army meant that German Communists were by no means a dead letter. The KPD in London was numerically larger than the SPD, consisting of 300 members as opposed to the SPD's 160.[9] Faced, for the first time in three years, with an opportunity for meaningful external work, the SPD was thrown into a quandary. It was preserved from what seemed its only logical fate by the territorial demands of Stalin's government. For, as the Red Army moved west after 1943, and as Stalin made his claims to German lands plain, and as the Western Allies proved unwilling or unable to contest them, so the SPD was able to make firm political capital out of the fact that German Communists blindly supported the USSR and were helping to dismember the nation they alleged they wanted to help. The SPD, on the other hand, could demonstrate that at grave risk, it had stood up for the true interests of the 'other' Germany. SPD leaders feared that co-operation with Communists during wartime would become a permanent arrangement. How could the German people accept co-operation during the war against Germany but not in the task of reconstructing it?

In June 1943, Vogel and Ollenhauer were approached by Professor R. Kuczynski, a respected scholar who was a known Communist.[10] Despite certain suspicions about the real reason for the talks with German Communists that Kuczynski proposed, it seems that the SPD leadership was expecting yet another Communist attempt at the formation of a popular front.[11] It is not possible to say whether the planned FDB

---

[8] See Matthias, op. cit. (1968), p. 238 and Edinger, op. cit.

[9] Röder, op. cit. (1969a), pp. 32, 47.

[10] SPD *Mappe* 150, 15 December 1943, Vogel to Rosenfeld.

[11] SPD *Mappe* 51, Heine letter of 26 June 1943.

had its immediate origins in events in Moscow or in London.
Röder's view is that local initiatives were responsible, most
notably the publication of Heinrich Fraenkel's book.[12] Fraenkel
himself, who must be regarded as the foremost propagandist
of the exile rather than its historian, suggests that a combi-
nation of events in London and in Moscow lay at the root of
the invitation.[13] At any rate the second of the two meetings
Ollenhauer and Vogel had with the Communists occurred a
few hours before Radio Moscow announced the formation of
the NKFD,[14] and since it is very likely that Communists in
Russia directed KPD policy in Britain, it would seem certain
that the KPD's proposals were inspired by a general policy
emanating from the USSR rather than London.

It soon became plain that what the KPD was suggesting was
something rather different from what had been expected. As
early as 20 May 1943 Siegbert Kahn, a leading member of the
KPD in London, had drawn up a set of 'ideas on the formation of a unity committee of German anti-Fascists in Britain',
based on resolutions said to have been made at the so-called
Westphalian peace conference of 1942.[15] The KPD argued
that the United Nations and the resistance movements in the
occupied territories were engaged in precisely the same struggle
as anti-Nazis inside Germany whose interests should be represented by a common committee based on the following
manifesto. There should be no compromise peace with the
Nazis; the British should make an 'all-out war effort'—or in
other words a second front in Europe ought to be established;
the existence of progressive forces inside Germany should be
recognized; a free and democratic republic should be established there to prevent any return to reaction; and borders
should be set up within the terms of the Atlantic Charter.

Vogel and Ollenhauer were now confronted with a most
serious situation. Not only did the KPD proposals go very far in
meeting all previous objections made by the SPD, but many of
the leading Social Democrats in exile were greatly taken by the
idea of a FDB. The most important of these was undoubtedly

[12] H. Fraenkel, *Germany's Road to Democracy* (London), 1943).
[13] Cf. H. Fraenkel, *Lebewohl Deutschland* (Hanover, 1960), p. 61 ff.
[14] FO 371.39119 c 7450.
[15] Also SPD *Mappe* 62, and Röder, op. cit. (1969a), p. 199.

Viktor Schiff who worked for the *Daily Herald* and was considered the most highly placed SPD member in British life. Indeed, Schiff was to resign from the London group of the SPD over its refusal to join the FDB. He took with him Rawitzki and Adele Schreiber-Krieger.[16] Other leading exiles, either SPD members or close to it, also joined the FDB, like Leopold Ullstein, Irmgard Litten, Henrich Fraenkel (who had done more than anyone to argue the existence of 'another Germany' in Britain), Otto Lehmann-Russbueldt (the revolutionary pacifist), Arthur Liebert, and August Weber, the last chairman of the *Deutsche Staatspartei*. The German Communists were represented by Hugo Gräf, Wilhelm Koenen, Karl Becker (who had all been Communist *Reichstag* deputies in the Weimar period), Kahle and Hans Fladung amongst others. Koenen, being a member of the central committee of the KPD was perhaps the highest ranking Communist in Britain.[17] The FDB was able to unite on the basis of a common belief that 'if all those who believed in the future of democracy in Germany were to amalgamate, they would be stronger'.[18] At the founding meeting of the FDB on 25 September 1943 at the Trinity Church Hall in London, attended by 400 people, a twenty-three member committee was elected of whom ten were members of the KPD and three fellow-travellers.[19] The Communist chairman Bathke appealed first to the Allies when he informed them that an 'antifascist opposition now had the solidarity to overthrow Hitler', that they accepted Germany's responsibility to make reparation and to start a revolution inside Germany if they could.[20] Bathke then spoke to the German people and warned them that the only way to preserve their national independence was to revolt against Hitler. This would also give them equal status with the other nations fighting Nazism. Messages of support for this movement were read out from an AEU branch (which later urged the Foreign Office to give the FDB official recognition),[21] Dame Elizabeth Cadbury, and the Editor of the *Evening Standard*.

[16] SPD A. *Mappe* 13.
[17] Cf. Röder, op. cit. (1969*a*), p. 208.
[18] FO 371.34416 c 1160, 8 June 1943.
[19] FO 371.34416 c 1160, Nov. 1943. Cf. Röder, op. cit. (1969*a*), p. 203 ff.
[20] FO 371.34416 c 11608.
[21] FO 371.34416 c 11608.

It is interesting to note that the manifesto of the NKFD was very similar to that of the FDB. It had stated that Germany had lost the war and the 'patriotic elements inside Germany' must now overthrow Hitler and form a government to negotiate for peace. The German army should be withdrawn to the border. There should be a full democratic 'programme' but the new Germany ought also to be 'strong enough' to suppress Nazism. Nazi victims were to be released at once and an amnesty was to be offered to all Nazis who joined the NKFD.[22] Leading non-Communist exiles had slightly differing views about the aims and the nature of the FDB. In the opinion of Heinrich Fraenkel, it was the lack of unity amongst democrats that had allowed Hitler to come to power. Co-operation with the Communists was possible and desirable on the basis of Stalin's distinction between Germans and Nazis and any recognition was better than none. Schiff on the other hand, arguably the most influential Socialist exile in London at the time, seems genuinely to have accepted the NKFD as a shadow government which would enable Germany to avoid a second Versailles.[23] For some time Schiff had believed that the Vansittartist element in the British Labour party would make it impossible for German Socialists to influence Allied peace policy and the best tactic therefore lay in an alliance with Moscow via the KPD.[24] Other members of the SPD shared these views and there can be little doubt that Ollenhauer and Vogel faced a major political crisis within the SPD which lasted for the next six months. They were forced to consent to individual members of the SPD joining the FDB (a proposal put forward by Hans Gottfurcht).[25]

The zenith of the FDB was reached on 13–14 November 1943 when a delegates' conference met in London. A central executive committee of eight was elected and it seems that Siegbert Kahn was nominated as general secretary with the responsibility of co-ordinating commissions on the press and propaganda, on prisoners of war and examination of BBC

[22] FO 371.34414 c 8488, Nov. 1943.
[23] H. Fraenkel, op. cit. (1960), p. 63 ff. Cf. 34413–3770, Ollenhauer to Stampfer, Jan. 1943.
[24] H. Fraenkel, op. cit. (1960), p. 63 ff. PRO 34413–3770, Ollenhauer to Stampfer, Jan. 1943.
[25] Cf. Röder, op. cit. (1969*a*), p. 206.

policy towards Germany. Local groups were set up in Leeds, Glasgow, Birmingham, and London.[26]

After the Teheran conference in November 1943, however, the deep structural flaws in the foundations of the FDB were beginning to make themselves apparent. Kahn had attempted to get the FDB to pass a motion of support for the results of the Teheran meeting.[27] Schiff, who was as uncompromising a supporter of German national integrity as Friedrich Stampfer, was incensed by these moves to support the territorial reorganization of the eastern part of Germany. He called the proposed cession of Danzig East Prussia and Silesia 'one of the greatest injustices in the history of the world' and a 'gift to Goebbels'. Unless the FDB immediately condemned such a policy, he would resign forthwith. In January 1944 after a heated meeting of the FDB, it was clear that unless a compromise was achieved, the body would disintegrate.

On 3 February 1944 Kahn decided to get support for a secret five point programme,[28] which he hoped would keep the FDB intact. There are two versions of this document extant with significant differences between them. It was agreed that Germany should be free, democratic, and independent, and (though this is missing from the version in the SPD archives) its frontiers were to be those of 1933. Second, it was agreed that any annexation of German territory would be considered a 'grave misfortune' (this, too, is missing from the second version) and that the Hitler clique was threatening German sovereignty. It was further agreed that responsibility for reparation, the punishment of war criminals, and guarantees against any repetition were to be provided by the German people, and finally that all this would depend on what the German people themselves did to get rid of Hitler, and that the FDB ought to aid them in this.

Yet as the war progressed, it was plain that German Socialists like Schiff could not claim to speak for German interests and at the same time continue to support a body which fully supported what were seen as Communist plans for the dismemberment of Germany. The Communist view—whether genuine or

[26] FO 371.34437 c 10609, Nov. 1943.
[27] FO 371.39119 c 7450, Jan. 1944.
[28] FO 371.39119 c 2101, Feb. 1944.

or contrived—was that Russian plans for the future of Germany represented a stern warning to the German people and that they would deserve nothing better if they failed to get rid of Hitler. Koenen argued more specifically that it was inevitable that part of Posen and Silesia would have to be ceded. But the FDB had entered its death throes. On 22 February 1944 Churchill stated that the principles of the Atlantic Charter could not apply to Germany. This was widely interpreted by the exiles as an attempt by the West to prevent the USSR from making a separate peace with Germany by offering Stalin territorial aggrandisement for the USSR at German expense.[29] Sixty left-wing Labour MPs voted against the British Prime Minister, and were attacked by German Communists with vigour.[30] In April Heinz Schmidt sent Beneš a letter on the occasion of the Red Army's crossing the Czech frontier in which he assured him of the FDB's awareness of the guilt and responsibility of the German people and of their 'shame that they permitted the enslavement of Czechoslovakia and in no way contributed to the shortening of the war'. This was followed by a counter-letter from Schiff, Weber, and Wolff who warned of the dangers of anti-Germanism.[31] Finally, in May 1944 the split could not be bridged, and Schiff, Weber, Fraenkel, Leopold Ullstein, and Irmgard Litten all left the FDB and it subsequently collapsed.[32]

The collapse of the FDB caused great relief among the leaders of the SPD in exile. The overtures of June and July 1943 and the stormy debate inside the London group had been the beginning of a period that had almost killed the SPD. The political initiative had been firmly grasped by the KPD and there was every indication that some of the authority and the legitimacy, which the SPD claimed for itself as the real representative of the opposition to Hitler, had been passed on to the FDB—and was never to be regained.

SPD policy appears to have consisted of a public attitude and a private one. The former can be seen in the various public utterances that the leadership made, as well as in the letters

[29] Cf. Vogel letter to Seidel, 7 March 1944, SPD *Mappe* 140.
[30] FO 371 c 39119—7450, March 1944.
[31] SPD *Mappe* 42, 9 May 1944.
[32] Cf. Fraenkel, op. cit. (1960), p. 70. Fraenkel claimed the FDB was 'neither free, nor German nor a movement'.

sent to rank and file party members throughout the world, and the latter in letters to leading Social Democrats in America and Sweden. Vogel and Ollenhauer took great pains to gain international SPD approval for their policies.

Publicly, the SPD leadership promised there would be no co-operation with the FDB because it mirrored the policies of the NKFD.[33] Stress was laid on its alleged bourgeois character and aims. It was said to be the marriage of those who had helped Hitler to power, namely the army, the capitalists, and the bourgeoisie, with those who had destroyed the German Labour movement from within, namely the Communists. In a speech made in London, Vogel argued that the NKFD did not represent the views of free Germans but was to be seen simply as a political pronouncement by the USSR.[34] Its manifesto was capitalistic. It contained no promise that the relationship between the property-owning classes and the means of production was to be altered. There was not a word about state control of industry and the break-up of landed estates in the east. Rather it was to be seen as a straightforward return to the politics of the Weimar Republic. Nazis and Communists were making yet another attempt to destroy the German Labour movement and the chance of creating a democracy in Germany. The SPD, Vogel concluded, would never co-operate with the FDB because it favoured the dismemberment of Germany. In addition it demanded that the German people accept collective guilt for Nazi crimes, and it would offer an amnesty only to those who had actively fought Hitler.[35]

The SPD's public attitude was summed up well in Ollenhauer's words to a rank and file party member: 'The new German democracy can only be constructed in a fight against Bismarck's grandchildren, never by fighting together with them.'[36]

Privately, however, the SPD was less confident. Heine had written to Vogel in October 1943 that 'a once and for all solution to the question of cooperating with the Communists is

[33] Ollenhauer to Stampfer, 10 October 1943, pointed out 'There are no generals here but Kahle was a lieutenant in the Spanish Civil War.' SPD *Mappe* 182.

[34] Vogel speech 1943, SPD *Mappe* 158.

[35] SPD *Mappe* 158.

[36] SPD *Mappe* 182, 8 September 1943, Ollenhauer to Reinbold. This was a policy advocated by the KPD at the height of the popular front period.

just not possible'.[37] In December 1943 Vogel wrote to Heinig in Sweden and on 9 February 1944 to Tarnow in Sweden that a 'union arrangement' with the KPD was thinkable and that a Weimar-style coalition with the KPD rather than the DDP or Centre party was 'possible and might be a great advantage.'[38],[39] Rawitzki was told in July 1943 that despite advantages for the SPD it might fall apart if it joined the FDB 'and it ought not to risk this for the sake of questionable cooperation with Communists and the completely apolitical Jewish and religious refugees'.[40] On 18 October 1943 Steuerwald was informed that 'the time for forming a Free German committee has still not come. We ought at least to wait for the meeting of the Big Three.'[41]

There was thus a difference between the caution expressed privately and the policy of non-involvement declaimed in public. The view that the NKFD and the FDB were simply bourgeois constructions which all true Socialists should ignore was ideologically defensible. In fact, it could be said to represent the clear view that there should be no repeat of the events of 1918—there should be no pact between Socialists and army leaders, no compromise over the issues of socialization and expropriation of the landowning classes. The opinion that the NKFD was simply a propaganda move by the USSR government was shared by the British Foreign Office, even if the Socialist leaders did not believe it sufficiently themselves to totally ignore the NKFD. But privately there was less conviction that the SPD would be able to maintain its stand and so stress was laid on the inappropriateness of the offer at that particular time.

Successful exile politics appeared to demand SPD participation in the FDB. If the SPD wanted political recognition, it had to come through membership in a comprehensive exile organization. The SPD knew that this was the case and it knew that many members desired it. The SPD's insistence that the real opposition to Hitler was working class and not bourgeois was moonshine. Privately, Socialist leaders knew there was no

[37] SPD *Mappe* 51, 6 October 1943, Heine to Vogel.
[38] SPD *Mappe* 150, 14 Dec. 1943, Vogel to Heinig.
[39] SPD *Mappe* 150, 7 Feb. 1944, Vogel to Tarnow.
[40] SPD *Mappe* 150, 8 July 1943, Vogel to Rawitzki.
[41] SPD *Mappe* 150, 18 Oct. 1943, Vogel to Steuerwald.

effective resistance to Hitler, effective in the sense that it could actually depose him, except for that in the German army.[42] Similarly, each month that passed without there being any organized opposition to Hitler, saw the 'good German' thesis buried more deeply in oblivion.[43] The SPD's leaders tried to put a brave face on their despair. As Ollenhauer wrote to Stampfer on 14 October 1943

the real decisions will not be taken by the *FDB* but by others, elsewhere —perhaps even at the Foreign Ministers' meeting in Moscow this month.[44]

He told Reinbold that although the SPD now had to work in a 'vacuum' it had at least preserved its freedom, something it could not do if its seat had been Moscow.[45] Hansen was informed that a 'free Socialist party would emerge after the war and that its most prominent leaders would be those who have gone through this party and have remained loyal to us in the years of dictatorship'.[46] The single most important motive for refusing to co-operate with the FDB however was the SPD's distrust of Communism. Ollenhauer had taken the dissolution of the Comintern as a move to encourage the western democracies into providing a second front in Europe rather than a pro-Socialist overture.[47] He argued that the German Communists could produce no coherent policy together with the Russians, because they would be torn between the concept that there was an important Communist resistance inside Germany and the Russian territorial ambitions, which were justified by the necessity for forcibly re-educating the German people.[48] The other factor in the SPD's attitude was its own political future. Behind closed doors, the Socialist leaders took pains to criticize every Allied policy which seemed to interfere with German rights of self-determination. An MI5 report on a meeting of the London party members on 18 June 1943 stated that

---

[42] In his letter to Stampfer of 14 July 1942 Ollenhauer had dismissed SPD talk of the existence of a pre-revolutionary situation in Germany, as a 'fairy tale', SPD *Mappe* 82.

[43] SPD *Mappe* 82, Ollenhauer to Stampfer.

[44] SPD *Mappe* 82, Ollenhauer to Stampfer.

[45] SPD *Mappe* 82, 8 Sept. 1943, letter to Reinbold.

[46] SPD *Mappe* 83, letter to Hansen, 7 Oct. 1943; SPD *Mappe* 82, 14 Oct 1943, letter to Stampfer.

[47] SPD *Mappe* 82, 6 June 1943, Ollenhauer to Stampfer.

[48] See Matthias, op cit. (1968), p. 609.

Vogel had spoken out very strongly against what he claimed were plans to 'dismember the Reich'.[49] He warned his listeners of the dangers that alleged American military occupation of Germany might bring. Its aim, he believed, would be to prevent a Social Democratic revolution from breaking out. Similar sentiments can be seen in the international policy statement by the SPD published on 23 October 1943, which was in part a response to activities of the FDB and an attempt to regain some of the political initiative in exile affairs.[50] There was nothing in this document that could possibly cause embarrassment to the Socialists who might emerge after the war, nothing that could be taken to imply any criticism of the German working class nor even of the bourgeoisie for that matter. The SPD claimed that the only way to ensure that German people would renounce the use of aggression in the future was that they be 'given the chance of following their own initiatives in shaping their internal political and social life'. Military occupation was not something to be desired and the cleansing of the Augean stables could be undertaken by the German people themselves working under the direction of a Social Democratic Hercules.

Such a statement did not in any way fulfil the political demands of 1943. But it would be politically useful for the SPD after the end of the war to be able to demonstrate that it had said such things. In addition, the SPD had been able to make most effective use of its anti-Communism. It had been able to avoid choosing destruction at the hands of the KPD in 1943 and 1944 by standing up for national policies based on the territorial integrity of the German nation. The Allied plans for Germany and Churchill's speech of February 1944 showed how wise the Social Democrats had been. Above all, as far as the internal development of the SPD in exile was concerned, the ideological orthodoxy of anti-Communism had been used in the best way. The lack of confidence, the confusion, and self-doubts that the FDB incident had engendered were explained away in terms of traditional party values. Inaction—which had almost brought about the complete disintegration of the SPD—could be justified in terms of strict party discipline

[49] FO 371.34414 c 8000.
[50] FO 371.34475 c 10609; see below p. 239.

and full obedience to the original mandate which had been given in 1933.

It is almost possible to sense the relief in Ollenhauer's remarks to Heinig in Sweden on 21 May 1944 in which he explained why he and his colleagues had done nothing to bring the SPD closer to the FDB. Necessity and virtue were truly united

There is really no sense in giving the impression through the creation of such an organisation, that we émigrés either can or indeed want to play at being normal political parties. The chance composition of the exile community and their lack of numerical and political significance exclude any such role. We in exile do not replace the party back home, we simply represent its ideas and act as its trustees.[51]

And, indeed, it was this role as trustees of the SPD that was to earn the London leadership such great respect at the end of the war and in the years thereafter.

[51] SPD *Mappe* 83.

XI

The Internal Development of the SPD 1943–1945

Now that it had failed to establish a reasonable working relationship with the Foreign Office and the Labour party, and
faced with a strong Communist onslaught, the SPD was forced
to rely more than ever on its internal strength. It was only this
that gives the plans and proposals that the SPD produced in
London a special importance. They show that German Social
Democracy refused to disappear and, more than this, that it
was still determined to change the face of future German politics through a mixture of political reform and social engineer-
ing. It goes almost without saying that in fact these plans were
never realized. The SPD did not gain power and by the time it
was in a position to direct German affairs, the time for radical
reform had apparently long passed. Yet two aspects of this
exile policy-making do require special emphasis. The first is
the extent to which Social Democrats in London were attuned
to the kind of ideas that were to help create the welfare state
in Britain after 1945.[1] The second is that they provide in-
controvertible evidence of the exiles' will for power during
this period.

As the war drew to an end, this will for power demonstrated
itself repeatedly. The London leadership demanded allegiance
from other Social Democratic exiles throughout the world.
Indeed, the primary purpose of the SPD's plans was to establish
the London leaders' supremacy after Hitler had been defeated.
Usually allegiance was forthcoming, but one very serious
struggle for power emerged in March 1945 when it became
obvious to Vogel and Ollenhauer that the SPD leaders in the

---

[1] See J. Harris, op. cit., p. 435: E. F. Schumacher who advised Ollenhauer also
advised Beveridge. He, together with Frank Pakenham (Lord Longford), Barbara
Wootton, and Nicholas Kaldor were members of B.'s 'technical committee'.

USA were determined to oppose the London leadership's bid for power. It was, without doubt, a highly significant episode.[2]

On 9 March 1945 Tarnow, an SPD leader in Sweden who was held in very high regard, wrote to Ollenhauer about the differences between the London leaders and the exiles in the USA.[3] He himself was most anxious to assure Vogel and Ollenhauer that his group would fully support them and were prepared to fight for their victory. But Tarnow was fearful that, unless firm action was taken by Ollenhauer, a number of conflicting Social Democratic heirs might emerge in post-war Germany. They should learn from the Italian example where after defeat 'every politician had wanted to be a political leader and so, overnight, forty political parties emerged'. This danger was that much greater, Tarnow believed, because the American exiles were determined not to return to Germany at once but to wait for the situation there to become clearer. This would give too many would-be SPD leaders a chance to cause trouble. William Sollman, who was not proposing to return himself in any case, had received a lot of support for his prediction that after Hitler German conditions were bound to be so catastrophic that the SPD should not allow itself to be associated with them in any way. The victors should be made to restore reasonable order first.

Vogel and Ollenhauer agreed with Tarnow and one week later they wrote to the SPD Executive members in exile in the USA, namely Juchacz, Aufhäuser, Dietrich, Hertz, Rinner, Sollmann, and Stampfer.[4] Despite the diplomatic language they used, their message was quite plain. If the American exiles could not agree to support the London group, they should dissolve themselves forthwith and cease any further independent political activity. Vogel and Ollenhauer argued that now rather than later was the right time to consider re-forming the SPD's Executive on a post-war basis to deal with German domestic politics in the future. The only real question that should be asked, they stated, was whether

we want to start again on the basis of differences, new and old, which would make reconstruction very difficult. Ever since those of us in London arrived in England, we were recognised by the British Labour

---

[2] Röder, op. cit. (1969a), p. 216, minimizes the importance of this struggle.
[3] SPD *Mappe* 185.   [4] SPD *Mappe* 13, 16 March 1945.

movement as the sole representative of the *Parteivorstand* and Stampfer gave us a kind of blanket authority to proceed . . . Although our activities here have been limited by factors which we do not intend to go into here, we have made very good use of our time, especially by uniting with the Social Democratic splinter parties, the *ISK*, the *SAP* and the *Neu Beginnen*.

Bearing in mind that the relations between the exiles and those in the Reich were an unknown quantity and that because of this the SPD leaders would only be able to assert the authority of their 1933 mandate 'for a short time', they must not only act but they must act at once.

Thus, in view of all the London group had done, Vogel and Ollenhauer now proposed that they should assume full responsibility for all the decisions affecting the SPD without waiting for support or listening to criticism from their American comrades. Furthermore, by pointing out that those in America had caused a lot of trouble before 1940, trouble which would be of no interest to the future Germany, Vogel and Ollenhauer were implying that even if the American group would not accept their bid, they would none the less pursue it. Indeed, they informed their colleagues that they were proposing to do two things at once. The first was to issue a proclamation to the German people and the second was to issue a formal claim to the entire property of the SPD. Although the Americans were not told this, Ollenhauer had, in fact, as early as 16 January 1945 asked a trusted SPD member, Max Lippmann, who was serving with the Allied forces in Paris, to ascertain whether the SPD's funds were still in the vaults of the bank where they had been deposited in 1940.[5] He was relieved to learn on 24 January 1945 that 'both money and documents are safe'.

Much to Vogel's and Ollenhauer's distress, their American comrades refused to comply with the demands laid down and agree to give up what authority they had. On 29 May 1945 Aufhäuser, Dietrich, Juchacz, and Hertz replied that although they accepted that the SPD should be reactivated, they did not believe that any members of the old Executive should play a part

Rather, it is the German workers, who maintained their Social Democratic traditions, who should discuss the future shape of the party. It is

[5] SPD *Mappe* 82.

wrong for you, Vogel and Ollenhauer, to base any claim on a mandate given twelve years ago under totally different circumstances.[6]

The Americans were prepared to assume the role of elder statesmen and they urged Vogel and Ollenhauer to do the same. But it would be quite improper for them to play a more active role in German politics or pretend they represented the continuity of German Social Democracy. Not surprisingly, when Vogel replied on 3 July 1945, he reacted with anger. He had, he said, expected them to be 'constructive' and he did not like their 'negative response'. Their mandates would not apply for ever but only until a new party Executive could be elected. But the fact was that very important political decisions would have to be taken before that could happen. Vogel noted that already some old SPD leaders in Berlin had re-emerged claiming, with no authority whatever, that they led the SPD. This would produce chaos unless the London group acted firmly. Above all, the Central Committee of the KPD, led by the 'old set' of Pieck and Ulbricht, had already produced a major policy declaration.[7]

The view that the Americans had expressed that Fritz Heine was the driving force behind the London group's decision was, Vogel added, quite ridiculous. Although Heine was 'bound to be one of the most valuable members of the new SPD', he was not their leader. With this observation all correspondence between London and America about this matter and, it would appear, about all other matters ceased. The Americans, many of whom did not wish to return themselves, had tried to prevent the Londoners from going back as the official leaders of the exiled SPD. This was something that Vogel and Ollenhauer would not stand for. It was a stroke of luck for the London leaders that in disputing the Londoners' claim the Americans did not advance one of their own.

In the event, of course, the Americans' argument was quite wrong. For better or for worse the SPD leadership in London had carefully staked a claim based on its past and present political work and this claim was respected in Europe. The ease with which Vogel, Ollenhauer, and their colleagues asserted their own authority during this crucial period is ample evidence

---

[6] SPD *Mappe* 13.
[7] See H. Krisch, op. cit.

that their policy had paid dividends. They were not forced to retreat but were able to stand up firmly for what they saw as the German Social Democratic interest. As the Nazi government of Germany, which had seemed so impregnable, first tottered and then collapsed, German Social Democrats in England began to prepare for what they believed was their hour. On 30 April 1945, two days after his mentor Mussolini had been executed by Italian freedom fighters, Hitler committed suicide in Berlin. On 4 May 1945 his successor, Doenitz, agreed to surrender the German forces in North Germany, Holland, and Denmark, then, on 7 May 1945, General Jodl for the German High Command surrendered to General Eisenhower and the war in Europe was over.

The British people celebrated this event on 8 May, VE day, the day when victory and the defeat of Nazism symbolized a new start in European politics. Ten days later, on 18 May 1945, the SPD leadership in London issued a declaration demanding the 'creation of a new, peaceful, Social Democratic Germany'.[8] Millions had died and those guilty for their deaths, who were identified not as the German people, but simply as 'the Nazis and the militarists' now needed to be punished. Although the Germans themselves had suffered under National Socialism, the SPD argued, they would have to bear *Mitverantwortung*, co-responsibility, since they had done nothing to 'actively hinder the regime'. In addition, they would have to 'cooperate loyally with the occupying authorities'. At the same time, however,

the new Social Democratic Germany cannot be created from the outside but only from within German politics in a struggle against the political and economic adversaries of the new state.

For this reason the SPD leadership urged that the German people be allowed 'an education in democracy', and that the SPD should therefore be allowed to begin its political work straight away aided by 'a free press' and Allied permission for its work and that of any other 'free political organisation'. No one should forget, the declaration concluded, that it had been the SPD who had first warned the world that Hitler wanted war and that the SPD had also 'played a leading role in the peace movement of 20 July 1944', a claim that was not wholly

[8] SPD *Mappe* 184.

borne out by the evidence.[9] In fact, although Social Democrats had played a part in the plot against Hitler, it was not a very significant part. But it is an important comment on the SPD at this time that it wished to capitalize on the July plot and to identify itself wholeheartedly with it.

Exile was clearly drawing to a close for the SPD leaders in London and they were now able to enact some of the concepts they had formed about their own political status. They were at last in a position to see whether all that they had worked for stood a chance of succeeding, whether the work they had done would actually permit them to act as the heirs of German Social Democracy, (though not the sole heirs), and whether they would be seen as the link, the vital link, between the past, the present, and the future. If chiliastic policies had always been part and parcel of Socialist ideas, the time was rapidly approaching when what had been planned for Germany's future might be put into effect. The most important concept they had formed was undoubtedly that they should occupy a leading position in the German Labour movement after Hitler. This concept was not only what had kept the SPD alive during the war, it was what had also alienated the party from the British Labour movement and, to a limited extent, from the British Foreign Office. Now, with each passing day, this concept required establishment inside Germany. There was not only the prospect that other SPD figures would try to seize the leadership of the movement, but that the KPD would, by virtue of its support from the Red Army, defeat the SPD in the battle for the political allegiance of the German working class before Vogel, Ollenhauer, and the rest could return. The London leaders spared no effort in trying to force the British authorities to accept their claim and restore them to their native politics by demonstrating their readiness to take over the direction of German political affairs and showing the Allies that they possessed realistic plans for what they would do with power.

In July 1945, for example, Social Democrats published a set of proposals for the 'immediate introduction of communal self-government'.[10] They demonstrate that the SPD was aware

[9] See above, p. 201.
[10] SPD *Mappe* 184.

of the fact that 'the creation of a new, free, democratic, and socialist Germany' would be held up by the limitations laid down by the armies of occupation but that in the meantime all Socialists would have to devote themselves to getting rid of Nazi influence. The SPD, therefore, urged that until fully democratic elections could be arranged, local councils or *Orsträte* should be set up composed of people of 'proven opposition to Nazism' namely workers, peasants, and exiles. These councils were to liquidate the Nazi party and imprison all local Nazi officials and every member of the SS, SA, Gestapo, and NSDAP members 'active in terrorist measures'. They were also to free political prisoners and set up adjudication committees *'Schiedstellen'*, to take over police affairs. No NSDAP members could be allowed to participate and only members of *Antifa (sic)* groups could actually work in the police force although those who had been police officials before 1933 and then dismissed could rejoin. The councils were to secure all food supplies and take them over, and to ensure that all workers were allowed to control their own works' councils. In order to avoid any unemployment, SPD leaders said, workers would have to help clear rubble.

The councils, the document went on, were to hand all printing activities and newspapers over to the 'democratic institutions' and also supervise the dismissal of all Nazi school teachers. Every district (*'Bezirk'*) was to have a court run by a lawyer to be nominated by the council. Finally, the entire population was to be scrutinized by tribunals which would possess the power to pass the death sentence and life imprisonment. Like all the previous plans outlined here it is possible on the basis of historical hindsight to dismiss it as engaging but of no practical import. Nevertheless, individual aspects of this proposal are, in themselves, of great interest. A recent work by L. Niethammer[11] places the emphasis on organizations inside Germany when discussing the chances of a fresh political beginning after 1945. The German *Antifa* groups, he argues, were the decisive spawning ground for fresh political ideas often developed in co-operation by Socialists and Communists. On the other hand, this document shows that when it came to radicalism the exiles should not be forgotten and that their

[11] See Niethammer, op. cit.

activities were already running parallel to those conducted by
their associates inside Germany. One important exception, of
course, was the co-operation with Communists. This is a central
and fundamental distinction between the SPD in exile and
Socialists and Communists in the field. The eventual outcome
of this conflict of opinion was quite obviously dependent on
a number of factors but one of these (and one that is often
ignored) was the attitude of the exiled SPD leadership to co-
operation with the KPD.

In September 1945,[12] all SPD members in London were
sent a circular by the leaders asking for their views on party
organization and outlining the views of the leaders themselves.
These were that the 'reconstruction of the party organisation
should begin at once' and that every member of a Socialist
organization including the *SAJ* and the *Rechsbanner*, should
be counted as a member. The aim should be to create a caucus
of 'reliable members' to protect the young organization from
Communist spies and *'Konjunktur Sozis'*. All independent
Socialist groupings before 1933 should be approached in order
to create a new united Socialist party on the basis of free and
democratic Socialism, that is with no possibility of any Communist collaboration. The organization of this new party
should be similarly unified, and it should be something more
than a mere working-party. This, it was pointed out, was a
central platform in the creation of union and one it wanted
perpetuated. Because ex-Socialists (*sic*) would provide the
majority of the membership it would be possible to invite
members of 'other organisations' to join the party. Whether
this was a veiled reference to Communists or to erstwhile
members of the Hitler Youth, for example, cannot be stated.

The document then went on to lay down some specific rules
of conduct which demonstrate the SPD's exile leaders' concern to remain at the head of any political movement inside
Germany to ensure that it did not get out of control especially,
it should be noted, with regard to possible collaboration with
Communists. Those 'decisions which affected matters beyond
the immediate locality should be left to a full party conference
to decide' and the programmatic principles and the guidelines
for the creation of democratic self-government 'should in all

[12] SPD *Mappe* 14.

cases be used as references and rule books'. The new party organization should be built 'as soon as possible' on internal party democracy and so all local groups ought to arrange to offer themselves for election. The 'highest aim' was to furnish the new representatives of the party as quickly as possible with the 'full authority of elected trustees of the party'. Because of the catastrophic conditions inside Germany, the document concluded, co-operation with 'other parties' might be necessary on a local level but the SPD must make absolutely certain that its 'organisational independence and its political independence and each single policy should not be compromised by any sort of cooperation whatsoever with the KPD'. Any suggestion of creating a single party (*'Einheitspartei'*) together with Communists (rather than an unified SPD *'einheitliche Partei'*) or even any 'local discussion of one' was to be rejected out of hand. The whole question of co-operation with Communists, it invited the SPD membership to recall, involved a very detailed examination of the goals and policies of such a party, of its relationship with democracy, and especially of internal party democracy—matters that local groups were not competent to assess.

At the same time the SPD issued an official plea to the Executive of the British Labour party.[13] The run of military defeats and the collapse of 'the Hitler dictatorship' meant that German Socialists 'must now face the task of rebuilding the SPD'. The comrades in Germany will 'depend a lot on the sympathy and support of Socialists in exile'. Having stated that they believed the London group would play a central role in the reconstruction of the SPD, they went on to ask for the help of the British Labour movement: 'the moral and material help that you have given us since our arrival encourages us now, and we are convinced that despite your own problems you are interested in the free and democratic workers' movement in Germany'.

The London leaders had 'no illusions about the difficulties they would face'. Germany had been 'politically, culturally and economically destroyed by the Nazis'. In addition there were specific party political problems—the defeat of the SPD in 1933, 'the old age of a lot of party members, 11 years of

[13] Ibid.

illegality and the existence of political apathy, and militaristic and nationalistic tendencies'. Yet precisely because of these difficulties, the SPD leaders had decided 'to rebuild the party speedily'. Factors made their success seem likely: news reports and the opinions of POWs all confirmed that the Social Democratic ideal still lived, there was still a considerable residue out of the 100,000 party officials and 900,000 party members. Furthermore, 'German politics in the first post-war period will be mainly influenced by the over-40's because of the heavy losses amongst younger groups, the effects of Nazification and so on.' Finally because 'of the cooperation of various Socialist groups in exile in England, the chances of creating a united Socialist party are better than ever before.' The German people needed to be politically organized as quickly as possible not least because opposed to the SPD exiled leaders there was the 'Moscow Committee' which was now being assisted 'with all the resources of a totalitarian state'. 'This black-white-red union of generals, militarists and Communists will have a fateful influence in a Germany that is occupied unless it is counterbalanced by a strong and free workers' movement.'

The London group concluded by asking that they be allowed to return immediately after the final surrender had taken place with an assurance that political activity would not be impeded. Secondly, they demanded that SPD property stolen by the Nazis be restored to them at once either by finding it and handing it over or by using Nazi assets as a compensation. Thirdly, they asked that the British Labour party send a representative with full powers to Germany together with the SPD exiles to co-ordinate all further efforts at co-operation. This sensible and urgent request appears not to have been answered. But the fundamental fact remains that the London group were now literally trying with every means at their very limited disposal to ensure they played a decisive role in the rebuilding of German Socialism. To reinforce this statement, Vogel (already seriously ill), Ollenhauer, and Gottfurcht went to see Morgan Phillips personally.

On 26 July 1945 the leader of the Labour party, Attlee, became Prime Minister and although the evidence of the past years did not make the SPD leaders believe they would get any special treatment or that their status would change overnight, they certainly hoped the new Labour government would not want to hinder the SPD's intentions towards the new Germany.

But even a friend of the SPD like Ellen Wilkinson, MP for Jarrow and Chairman of the Labour party for 1945, had to tread carefully when trying to expound the thesis that severity towards Germans might not be the best means of achieving democracy. In her chairman's address in May 1945, at the end of her period in office (she was replaced by Harold Laski) she stated that

We shall have to be stern to beaten enemies and keep old wounds open just because we must bring home to the Germans that a nation *is* responsible for its own government. A nation cannot escape responsibility by saying 'We did not know' or 'We could do nothing about it'.

Yet having said this she went on to argue that

The Labour party is particularly fitted to deal with the problem of how to find in Germany the people who can provide some kind of democratic leadership in the day to day administration of Germany under Allied supervision. After the First World War a tragic mistake was made. The democratic political heads of the Weimar Republic were treated as though they were themselves war criminals.[14]

Only a keen student of history might be expected to know that these political leaders were, of course, Social Democrats, or that the SPD might provide trustworthy administrators.

Whilst all this was happening the SPD leaders in London were beginning to receive the first reports from inside Germany. SPD members in France and Belgium had already resurfaced, most notably perhaps Heinz Kühn who on 3 July 1945 wrote to Ollenhauer that he had been hiding in France and Belgium during the occupation and that he now wished to represent the SPD in those countries. This was quickly agreed in London and he began to receive official papers. Kühn, it will be recalled, rose to very high office in the Federal Republic and, until the summer of 1978, he was Prime Minister of West Germany's most populous state, Nordrhein-Westfalen. Similarly Erich Brost, who had been sent to Germany by the British authorities, was invited to send full accounts of all that he saw. On 14 May 1945 he advised Ollenhauer to drop demands

[14] Report of the Labour party Conference for 1945, pp. 78, 79.

for the punishment of war criminals and German disarmament:

they find no support amongst the Germans. The people in Britain who like hearing them, that small group in the Labour party, refuse to listen and those who look deeper and are ready to seek German help in the future have already realised that a German political party that supports such things will have a very difficult political future. We might remain boring exiles if we don't change our tone and we must always be very careful to avoid the role of so-called good Germans. Harmless people are not now liked.[15]

It was Brost too who sent Ollenhauer the first report on Adenauer. On 3 August 1945 he wrote,

I have just returned from a very frank and long discussion with Adenauer. He wants to create a large *Sammlungspartei*, a widely based peoples' party from the conservative right to the positively minded (*positiv einge-stellten*) in the SPD.

The most important thing, he said, was that he had been visited on 2 August by a number of high-ranking American officials who asked him to recommend names for a Reich government. He suggested that they make Brüning the leader but they dismissed this as impractical.

Adenauer told me that it was made quite clear to him that if he wanted to he could play a very influential role in the new government but his problem has not changed since the Weimar days—he does not want to be *Reichs* Chancellor because he likes living in Cologne.

Adenauer's solution to that particular problem was original and effective but it is interesting to note that in his memoirs he gives no account of this meeting with American officials. It is by no means insignificant however, that Adenauer relates that at this time he had a lot of trouble with an erstwhile exiled SPD member by the name of Görlinger who had allegedly informed against Adenauer to the British Secret Service.[16] It was not contrary to Adenauer's habits to use the SPD leaders' exile against them in order to stress the 'very close relationship that the British Labour party had to the SPD'.

With each passing day the London leadership became even more anxious to return to Germany. The reports that were coming through indicated quite clearly that domestic German political activity was beginning to intensify and that unless the

---

[15] SPD *Mappe* 13.
[16] See K. Adenauer, *Erinnerungern 1945–1953* (Frankfurt, 1967), p. 22 ff.

exiles returned fairly shortly, they might not only be unable to participate in fundamental policy matters but also might lose their claim to leadership. Ernst Reuter was beginning to gain support in Berlin, Wilhelm Kaisen had been made Mayor of Bremen, and Otto Grotewohl (also in Berlin) was not only demanding allegiance from SPD members there but also calling himself 'chairman of the party Executive' and, it was alleged, was keen to come to some agreement with the KPD leaders.[17] It was plain that any SPD negotiations with the German Communists might prove binding even if the official leadership had not played any part in them or given them any official sanction.

Above all, however, towards the end of August 1945 Vogel and Ollenhauer were told that a Dr K. Schumacher was getting a name for himself in Social Democratic circles as a brilliant political organizer and potential future leader. Indeed, on 20 August 1945 Willi Eichler went to Hanover to speak to Schumacher and was evidently so impressed that he accepted him as actual leader of the SPD.[18] On the basis of what Schumacher had told him, Eichler wrote to Vogel, he found no difficulty in allowing the ISK which had merged with the SPD in exile in 1941 to continue to consider itself part of the SPD, a statement which implies that it was for Schumacher rather than Vogel or Ollenhauer to determine the precise direction the resurrected SPD would take.

Although Vogel and Ollenhauer may have been considerably relieved by the intelligence that Schumacher 'recognised' their mandate, they can not have been delighted with Eichler's indication that Schumacher would not await their return to summon a conference to German Social Democrats in order to start resurrecting the party. Schumacher argued that events in Berlin, where amalgamation between KPD and SPD was proceeding apace, could be taken as a 'precedent which he does not wish to see'. He suggested that Vogel, and 'possibly Ollenhauer and Schoettle', might want to come to such a conference which he wanted to hold in October 1945. It was obviously very important for the London leaders to participate

---

[17] SPD *Mappe* 35, 23 Aug. 1945. A youthful Willy Brandt was also being talked about in SPD circles for having ambitions and a reckless desire to make a name for himself. SPD *Mappe* 58.

[18] SPD *Mappe* 35, 23 Aug. 1945. See below, p. 207 ff.

in such a conference in order to retain their positions in the years to come.

Early in September SPD members in London were told by Ollenhauer that 'the most important work in preparing the reorganisation of the party in Germany is now being done by Dr K. Schumacher, previously of Stuttgart, now of Hanover'.[19] And, now that Vogel was ill, Ollenhauer wrote to Brost that although the London exiles were ready to help Schumacher to rebuild the party as soon as possible, 'he must act with our agreement and in personal contact with us'.[20] When Schumacher wrote that he was planning a conference for the first week in October, Ollenhauer and Heine made strenuous efforts to ensure they could be present. On 6 September Ollenhauer wrote to Noel-Baker who was now the Secretary of State for Air in Attlee's government to seek his help in getting to Germany. He argued that the attendance of the London group was vital for the 'democratisation of Germany and for the peace policy of the British Labour movement' and that there could not be a 'new free Social Democratic movement in Germany' without them.[21] Furthermore, Ollenhauer argued, it would be in the best interests of all concerned if the British contingent was allowed to travel in civilian uniform, implying that the London leaders did not wish, now that the fighting was over, to be seen to be the one thing they had tried so hard to avoid, namely the agents of the British. Noel-Baker took some time to make up his mind, for as late as 23 September 1945 Schumacher was told by Ollenhauer that he was still uncertain whether they would be allowed to come. But official permission did then come and Ollenhauer and Heine were able to go to Wennigsen near Hanover for the all-important conference on 5, 6, and 7 October.[22]

Ollenhauer and Heine prepared a statement to take with them to Hanover. Its purpose was twofold. First, to make sure that the platform of the resurrected SPD bore the imprint of the exiles' ideals and, secondly, to provide the continuity in German Social Democracy that would allow the new party to emerge from the old and vindicate the actions of the exiled

[19] SPD *Mappe* 35, 8 Sept. 1945.
[20] SPD *Mappe* 82, 4 Sept. 1945.
[21] SPD *Mappe* 141.   [22] SPD *Mappe* 14, October 1945.

leadership since 1941.[23] The London leaders were anxious to show that their exile was in no way a disqualification from future German politics. Hitler had been defeated by the Allies and not by the SPD. Hence contact with the Allies, in other words, exile itself, was part of German political reality

the rebirth of the party is taking place under new conditions because we have not won back our organisational and democratic freedoms solely by our own efforts. The mission of the party Executive in exile is coming to a conclusion with our return to legality.

In addition, the London leadership argued in favour of an SPD which would purge Germany of Nazism and militarism and make reparation for all the damage the Nazis had caused. The keynote of the party had to be peace abroad and Social Democracy at home—*'Friedenspolitik nach Aussen, SPD Politik nach Innen'*. It was too early to produce precise programmatic details but the SPD had to become different from its former self. For a start it needed to broaden its electoral appeal to become a 'collecting vessel, a *Sammelbecken*' for people of all classes. Every Socialist who believed in free and democratic Socialism would be accepted into the party but no contact whatsoever with the KPD could be supported. Ollenhauer wanted this point to be absolutely plain. This was in part because he had just received a disturbing letter from Erich Brost in which he wrote that many Socialists in the Western Zone appeared to support the creation of a Social Unity party or SPD in the West 'because it would fight for equality, exist in the whole of Germany and it would fight Nazism'.[22]

All the points were fully accepted by Schumacher and incorporated into the new party platform. Fritz Heine recalled that there was genuine amazement at Wennigsen when it was discovered that all the London groups' notions were fully acceptable to those Social Democrats who had remained inside the Reich.[25] It comes as no surprise to see that Ollenhauer himself was elected in May 1946 to the deputy leadership of the party, that Heine became Schumacher's closest aide, that Heine, Ollenhauer, Schoettle, Eichler, and von Knoeringen were all elected to the *Parteivorstand* as part of the select

[23] SPD *Mappe* 13, See Röder, op. cit. (1969a), p. 246.
[24] SPD *Mappe* 13, 16 Sept. 1945.
[25] Interview with Fritz Heine, 21 Aug. 1975.

salaried inner-circle members and that in 1952 Ollenhauer was elected leader of the SPD on Schumacher's death.[26] The bounds of the acceptance of the exiles' policies may even extend to the adoption of the Godesberger *Programm* of 1959, written largely by Eichler himself in which the SPD first openly committed itself to being a *Sammelbecken* for Germans of all classes.

Exile was ending. The London leadership had fulfilled its historical and its political mission. Its return to legal political activity in the new Germany could only be a matter of months. The only tasks that remained were concerned with organizing their final return. Many SPD members, perhaps twenty per cent, expressed reservations about returning despite the challenge of the future. Those who did return, however, were able to form an elite which did much to make western part of the new Germany a free democracy. They took up positions in a variety of professions, not simply in politics but also in administration, where there was a shortage of suitably qualified non-Nazis, in the arts and in the media.[27]

On 29 December 1945 Ollenhauer delivered his last speech in London. With Hans Vogel's death, which had occurred at precisely the same time that the SPD was resurrected at Wennigsen, Ollenhauer had been elected leader of the exiled Executive. He now recalled that five years previously, exiled Social Democrats had come together to form a union in order to achieve three major tasks. First, to defeat Hitler, secondly to represent Social Democratic policies and ideals, and thirdly, to prepare the ground for the emergence of a reborn SPD, *'die Wiedererrichtung einer neuen deutschen sozialdemokratischen Arbeiterbewegung vorzubereiten'*.[28] All of this had to be done despite the years of internecine feuding that had preceded exile in England, he added. Had they, he asked, been successful?

The answer is both 'yes' and 'no'. We were not able to realise our aim of getting a unified German Social Democratic party recognised as an equal partner in the fight against Hitler because of the potency of Vansittartism. In the second half of the war, the Allies made policies purely on the basis

[26] See H. Hermsdorf *Erich Ollenhauer, Mensch und Leistung, Ansprache zum 10.Todestag* (Bonn, 1974).
[27] SPD *Mappe* 141.
[28] SPD *Mappe* 12.

of *Machtpolitik* and the political and moral weight of the SPD proved too slight to counter this.

The SPD, Ollenhauer went on, had never claimed that there would be an internal Socialist revolution inside Germany although it did believe that there was within the Reich an indestructible Social Democratic opposition. Events, he alleged, had proved this to be the case and the SPD had been vindicated by the 20 July 1944 and the spontaneous re-establishment of a workers' movement on the ruins of the Third Reich.

At the same time, Ollenhauer, argued, the real need had been for a well-thought out programme or manifesto which a new unified Social Democratic party might follow. This had been done in London in three years, 'three of the most fruitful years of the SPD's exile'.

We in London have, then, achieved our highest aim, namely to make ourselves superfluous. The *Heimat* was always the source of our inspiration. Our mission as political exiles has now come to an end. But we have a new task, to make free Social Democracy victorious in Germany in order to prevent a new catastrophe. We must not waver in our ideals nor underestimate the difficulties nor even believe that socialisation of state and economy by itself is an end to aspire to. It is simply the means to set free the human spirit and lead it forward to a new era.

With these stirring words, the SPD in exile ceased to exist. There can be little doubt that Ollenhauer's estimation of the significance and the achievement of exile was correct. The SPD had been preserved as an independent political entity which had been able to maintain both an inner and outer existence. As the ideological basis of the Second World War changed, when it ceased to be primarily a war for ideals and became instead a war between nations, the external existence of the SPD was almost extinguished. Yet the other side of this coin, which was to threaten to destroy the national integrity of the German state and to create the conditions for the division of Germany, helped to offer new life to Social Democracy. It became the opponent of all those who wished to dismember Germany and deliver parts of it to totalitarian politicians who were, in the eyes of the SPD, as dangerous as the Nazis. Anti-Communism, which had since 1917 been an article of faith for the SPD, became a political lifeline, preserving the party in its darkest days and providing it with a platform which could transcend the negative implications of exile, of spending the Second World War in a country which was fighting to destroy the German government, of actually helping the defeat of that government by working for the Allied Forces.

Conclusion

THE political history of the SPD in exile in London has many themes. They range from the isolation, 'the emptiness of exile', as Hans Vogel put it, to the single-minded and courageous dedication of a group of professional politicians to the pursuit of power. Exile is in itself a most curious political phenomenon. It can lead to domestic political success but it can also lead to oblivion. The strength of Socialism and the strength of its chiliastic ideology explain why exile in the case of the SPD did not lead to oblivion. And in internal party terms it did not lead to failure either. But success for the SPD was qualified. It had assumed that it would follow Hitler and the Nazis to political supremacy but in fact it took eighteen years —until 1966—to achieve even an element of Federal control in post-war Germany. One reason for this is undoubtedly the fact that exile in England did not bring the sort of co-operation with the British government and the British Labour movement that the SPD leadership had sought. Had the exiles been accepted as independent equals and had the Allies wanted to see a united Social Democratic German Reich in 1945, the success of the exiles might have been complete.

Matters did not turn out this way, however. Perhaps it was unrealistic to expect that the war would continue to be fought on the basis of the ideas which led Britain and France to declare war on Germany in 1939. In any case without the nationalist concerns of the leaders of the Allied nations which led logically to their preferring agreement at the expense of German national integrity to disagreement at the expense of their own, the SPD would by 1944 have been in very serious isolation. All the same, it was not without irony that the SPD became the champions of a German nation which had been constructed by Bismarck and which had been led by those hostile to Social Democracy for far longer than by Social Democrats. Although this allowed the SPD to speak for the national interest as opposed to the international one, and although it made it almost impossible for the conservative opponents of the SPD to question its national identity, the cause of Social Democracy may not have been furthered. Certainly the SPD leadership was determined to stay in the business of political power. Its greatest fear during exile was, in the words of one of its members

that between those in the Reich and those in exile a measure of alienation, *Entfremdung*, would have taken place which might have ruptured the old feelings of solidarity and the political strength that emanates from it.[1]

As we have seen this did not take place and more than this the resurrected SPD was one close to the hearts and minds of the exiles. As Ollenhauer wrote, two important victories had been gained by the exiles. First their own position in the new party was assured. Secondly, the new SPD was not only going to be firmly anti-Communist. It would also oppose 'all political dogma. It will be a real peoples' movement, with a clear national policy. It will appeal to all classes.'[2] Above all, the mistakes of 1918 were not to be repeated. There simply was no division within the official, the real SPD in the Western part of Germany. And, just as importantly, the new Republic that emerged in 1948 had an opposition that was nationally minded, democratic, and loyal to parliamentary values.

This gave the Federal Republic and its Conservative leadership an advantage denied to Weimar and those who had created it thirty years earlier. It was, of course, a paradox that what the SPD did for the Federal Republic was in part done by being in opposition. But if a trustworthy opposition is a vital component of parliamentary democracy, then the SPD's role is not insignificant. Exile, then, kept German Social Democracy alive thanks to the efforts of brave and honourable men, men whose story demonstrates that National Socialism was not the only logical outcome of German political development,

[1] H. Gottfucht, SPD *Mappe* 45, 4 Nov. 1945.
[2] SPD *Mappe* 80. 2 Feb. 1946, letter from E. Ollenhauer to Katz in the USA.

even if for a variety of reasons it was, for a time, the dominant one. Under free democratic conditions the SPD flourished and subsequent years confirmed what its exile history had established, that as a people the Germans were no less capable than any other of wanting democracy and social justice. It was this knowledge that inspired the exiles in London and, ultimately, brought them home to Germany.

Appendix

Membership of the International Subcommittee of the Labour party during the period covered by this book
George Dallas MP (Belper), Hugh Dalton MP (Bishop's Auckland), Harold Laski (Central Fulham DLP), Rt. Hon. Herbert Morrison, MP (London Lab. party), Philip Noel-Baker, MP (Derby DLP), E. Shinwell. These figures remained on the committee broadly speaking for the duration of the war, with the addition of Mrs B. Ayrton-Gould, MP (Seaham DLP).

The Secretary of the Party was J. S. (Jim) Middleton, who retained this post until 1944 when he was replaced by Morgan Phillips.

Gillies's name as Secretary of the International Department and Secretary of the International Subcommittee of the NEC appears each year until 1945 when no name is given. In 1946 Denis Healey is shown as the International Secretary.

The Rt. Hon. C. R. Attlee was leader of the Labour party throughout this period.

The chairman and vice-chairman of the Labour party's NEC during this period were as follows:
1940-1, James Walker, MP (who died early in 1945) and W. H. Green, MP
1941-2, W. H. Green, MP and A. J. Dobbs
1942-3, A. J. Dobbs and G. Ridley, MP
1943-4, G. Ridley, MP and Ellen Wilkinson, MP
1944-5, Ellen Wilkinson, MP and Professor Harold Laski
1945-6, Professor Harold Laski and Rt. Hon. P. Noel-Baker, MP

Members of the Executive of the SPD in Paris in 1938
Otto Wels, Hans Vogel, Erich Ollenhauer, Siegfried Crummenerl, Erich Rinner, Georg Dietrich, Rudi Breitscheid, Rudi Hilferding, Wilhelm Sollmann, Friedrich Stampfer, Curt Geyer, Fritz Heine, Paul Hertz, and Marie Juchacz.

SPD leaders in London during the Second World War
Hans Vogel, Erich Ollenhauer, Curt Geyer, Fritz Heine, (Executive members); also Hans Gottfurcht, Willi Eichler, Erwin Schoettle, and Wilhelm Sander.

SPD leaders in the United States of America during the Second World War
Dietrich, Juchacz, Hertz and Aufhäuser (who had, in Prague, been expelled for a time), Rinner, Sollmann, and Stampfer.

Bibliographical Essay

THE sources for writing the above history are widely dispersed. One study, by Werner Röder, is repeatedly referred to in the text and footnotes. This was the first attempt to produce a descriptive account of the SPD and its splinter groups in London. But it contains no real analysis of some of the central features of the SPD's exile history and the fact that in exile the SPD existed in relation to the British Labour movement and the Foreign Office and not simply for itself. Röder makes no mention of British policies towards German political exiles, and a proper estimation of the SPD's political development and the quality of its leadership is almost wholly lacking from his work.

There have been a number of studies concerned with the history of those who fled Hitler's Third Reich. Some have dealt with the refugees from racial persecution and others with the victims of political persecution (although the distinction between them is not always as clear as it might seem). Of the studies of political exiles, several have dealt with the German Social Democratic Party (SPD) either in passing, or in greater detail. Usually these have consisted of two differing kinds of investigation, one a broad and descriptive account of the personalities and organizations that sprang up in exile, the other an excursion into the occasionally obscure realm of the 'Other Germany'. A number of opponents to the Nazi state alleged that another non-Nazi Germany existed of which refugees from Hitler were the representatives. However dubious such a claim may seem, it should not be forgotten that we still know comparatively little about the opposition to National Socialism. The opportunities for resistance in totalitarian regimes are limited and so it is certainly quite reasonable to suppose that there existed a political but non-military opposition to Hitler which has been overlooked and underestimated.

An analysis of these German opponents to Hitler is a matter of special importance and complexity because the horrors of the Third Reich have lent an important role to those who disagreed with Nazi political values. The apparent political strength of the Nazis and their apparently total command of the German people from 1933 to 1945 had an obvious effect on anti-Nazi actions. In addition to purely historical difficulties and technical problems such as the dispersal of documentary evidence, the consideration of this matter also raises overt political difficulties of which some even possess an element of topicality. Is it, for example, correct to speak of a political *party* in exile? Why can prominence in exile lead to prominence after return? Were not Social Democrats who opposed Hitler, as their opponents after 1945 sometimes alleged, guilty of treason against Germany?

Finally, it needs to be considered whether exile as a political experience can be related not simply to the kind of politics that was pursued but also the way in which policy was made. The significance of exile takes on an entirely different form if it is measured not in terms of whether exile policies were effective during exile or whether they were enacted thereafter but in terms of the success of the individuals in gaining political power through exile and exile policies even if at the time they seemed wholly ineffectual. In other words, exile as a political activity may have rules which differ from those of normal political activities.

To a greater or lesser degree most existing studies of the exiled opponents of the Third Reich have provided no satisfactory explanation of these matters. One of the first scholars to apply himself to the SPD in exile was Erich Matthias, whose work *Sozialdemokratie und Nation* was published in 1952. Based almost wholly on an evaluation of the SPD's or the Sopade's (as it liked to call itself at this time) publications in Prague from 1933 to 1938, it probed the origins of the SPD's defeat by Hitler and it speculated on whether the SPD had learned any lessons from its fate. Broadly speaking, Matthias concluded that the SPD had not been able to oppose the Nazis effectively because it had proved incapable of devising policies based on a constructive conception of nationalism. It therefore had nothing with which to appeal to the German national interest after 1929.

Matthias claims that an examination of the SPD's role from the time of Bismarck to 1933 which includes, of course, the SPD's policies during the revolution of 1918 and the Weimar era, shows that the SPD believed that 'nationalist' politics were simply 'democratic' politics and nothing more than that. The experience of exile, however, forced the party to think more constructively about the fundamental need for German nationalism and the SPD began, for example, to argue that Hitler's policies were not truly nationalistic because they lacked humanistic values. Out of this sort of thinking there emerged a new and more positive approach on the part of the SPD towards Germany and towards a new Europe.

As this present study shows, there is indeed some truth in this theory, for exile did produce a new attitude towards the question of German nationhood, and the origins of the SPD's post-1945 nationalism can be located in exile. Yet the reasons for this are far more complex and revealing than those adduced by Matthias. There is no little irony implicit in the fact that having been deposed by a German nationalist, SPD leaders in exile should decide to support German nationalism and this necessarily involved fundamental rethinking. At the same time, it must be noted, the SPD's traditional view of Germany's national interest was not quite as facile as Matthias seems to believe. Its Weimar role was less blameworthy than he implies. Unlike many of those who opposed it, like the *Stahlhelm*, the DNVP, the DVP, and of course the NSDAP, the SPD believed that the German nation ought not to pursue aggressive policies, a view that it incidentally did not abandon after 1945. It seems impossible to believe that had the SPD insisted less on fighting for a democratic Republic, the true interests of the German nation would have been better served. Finally, it should not be ignored that the SPD was always ready to be a loyal opposition in the liberal Weimar Republic which is a great deal more than can be said of those parties who claimed to be more nationally minded. Even so it was not hard for the SPD's opponents in Weimar to interpret its internationalism as anti-nationalism.

Matthias was convinced that he had discovered a new direction in the SPD's policies which had been produced by its exile experience in Prague and he called it a 'new orientation', one that underlay the development of all its ideas at this time. Some years after this work appeared, however, he seemed to have altered his view somewhat. In his edition of Friedrich Stampfer's *Mit dem Gesicht nach Deutschland* papers he argued that that exile had produced far less change in the SPD's fundamental political thinking. All it had achieved was to enable the SPD to distinguish more plainly between Communism and Socialism and to create some preconditions for post-war success by forcing the various Socialist splinter groups in exile to come together in London in 1941. Finally, the *émigré* Socialists had helped to give the SPD a decent reputation in the post-1945 period even though they had no real political influence in the subsequent history of the party.

The links between the exile period and the post-war years were seen as a decisive problem by a number of authors. Albrecht Kaden in *Einheit oder Freiheit* (Hanover, 1964) considers what he terms 'the problem of continuity in the SPD'. He demonstrates that the leaders of the SPD in exile had ideas surprisingly similar to those held by Kurt Schumacher who had emerged from within Germany to become the party's first post-war leader. Kaden highlights the SPD's outright hostility towards Communism in this argument as well as the party's strong wish to extend the basis of its support by rejecting a solely working class constituency for itself. This, of course, came to fruition in the Godesberger *Programm* of 1959 written in large part by Willi Eichler, who had been an exile leader in London (pp. 92–3). Furthermore, Kaden points out that many exiles who had been in London in the war assumed leading posts in the party after 1945. These included Fritz Heine, Herta Gotthelf, Richard Loewenthal and, of course, Erich Ollenhauer who became deputy leader of the SPD under Schumacher and then in 1953 succeeded him until 1961. All this, Kaden argues, must prove the existence of continuity within the party and as a consequence he concludes that 'neither externally nor internally was the SPD founded afresh in 1945, rather it was refounded', since the main support for the revitalized party came from the rank and file functionaries who had stayed in Germany from 1933 to 1945 (p. 281).

A very similar thesis is offered by Harold Kent Schellinger in his interesting monograph on the SPD in the Bonn Republic, *The SPD in the Bonn Republic* (Den Haag, 1968). He argues that the Second World War gave German Socialism a chance to reappraise itself and the fruits of this reappraisal can be seen in the process of modernization of the party and its doctrines which culminated in the 1959 manifesto.

But another scholar, Theo Pirker puts forward very different views, *Die SPD nach Hitler* (Munich, 1965). He states that exile provided the SPD with no new orientation and that this period had an absolutely minimal amount to contribute to the post-war development of the party (p. 9). Those in the SPD who put forward this view were, he suggested, simply trying to manufacture a continuity in the party's history in order to create a legendary past which absolved the SPD for responsibility for Hitler's success (p. 8). The exile period was insignificant, he concludes and the exiled political leadership brought 'no political ideas back (to Germany) with them save a naïve faith in the victorious British Labour party comrades' (p. 37).

Pirker's case is not substantially different from that made in one of the first descriptive histories of the SPD in exile which was written by Lewis Edinger, *German Exile Politics* (California, 1956). Despite the fact that he believed the SPD deserved credit for having been the sole organized opponent to totalitarian rule in Germany (a formulation which permits the exclusion of the German Communist party, the KPD), and despite the fact that it could have given Germany a 'succession élite untainted by Nazism', the actual history of the exiled Social Democrats was a tale of 'failure, frustration, of shattered hopes and bitter strife' (p. 247). They had underestimated Hitler, they failed to create an internal resistance organization and they could, in the final analysis, not prevent their being defeated by the Nazis because of their refusal to become revolutionaries (p. 247).

The most recent descriptive history of the SPD in London during the Second World War does nothing to alter this view, even though it takes pains to show the exiles to their best advantage *Die deutschen Sozialistischen Exilgruppen in Grossbritannien* (Hanover, 1968). Röder's account offers little to modify Edinger's conclusions despite a far fuller examination of the period from 1941 to 1945. Even though eight out of the thirty Executive members of the SPD in 1954 had been exiled in Britain and although seventeen per cent of the SPD *Bundestag Fraktion* in 1949 and in 1953 had been exiles, the only real results of the Social Democrats' existence in London were failure, frustration, and strife (p. 246). The most positive remark he feels able to make is that 'in the final analysis, exile and fatherland may be seen as part of the common experience of Social Democratic history and politics' (p. 247).

It is easy to be disappointed by Röder's account, and also by many of the others mentioned here. This is not because there is no consensus on whether exile proved a political success for the SPD. It is in any case impossible to speak of success or even of significance in conventional terms when exile politics are under scrutiny. Realistic success in exile might have consisted of far more modest achievements than success in domestic politics. To imagine that the SPD or indeed any exile organization could by itself and without force of arms have overthrown a totalitarian dictatorship as powerfully entrenched as Hitler's is absurd. However strong the SPD may have been during the Weimar era, its enemies had proved stronger. Success in exile, therefore, had more to do with sheer survival than whether the German people listened to the SPD's exhortations to overthrow their Fuehrer or whether any individual proposal for the post-war period was rejected or accepted. Indeed unless the need for survival is put into a prominent position, any analysis of the ideas the Social Democrats developed in exile—whether of a day to day variety or of a programmatic nature—will have a most limited meaning.

Furthermore to work solely on the basis of the SPD's documents from exile as most, if not all, of the above authors have done, is to provide a very one-sided and insubstantial account. Exile success was not simply about the internal state of the SPD but also about its relationship with other bodies and institutions and first and foremost with British ones. In the same way, the question of continuity or tradition in the SPD from 1933 through to the Bonn years has been dealt with in a glib manner. What does Kaden really mean when he speaks of the 're-founding of the SPD' in 1945?

The anti-Communism of the SPD after 1945 is but one example of a traditional SPD policy being revitalized to great political effect through exile and then being employed in a different objective situation with possibly even greater effect during the Cold War. At the same time, the political success of exile should not be measured in terms of ideological consistency. A change in policy (like the post-1945 view that the SPD should no longer seek to represent the working class only) might well be as significant as a consistent one (like the SPD's anti-Communism). Exile was less about continuity or change than it was about staying in the business of politics, and as much about the *ability* to formulate policy as about the policy itself.

Any sensible analysis of the political history of the SPD in exile ought not, then, to disregard these basic questions or shy away from offering answers to others which the existing studies have ignored. The most rarely asked question of all is perhaps the most radical: can the SPD really be said to have existed after 1933 and before 1945? In one sense it seems ridiculous to suggest that a political party can exist without a membership. Even if one can show that in exile the SPD could claim to have several hundred members who were also in exile, does this then make it a more plausible party?

The SPD in exile was clearly not the same sort of political party as it had been in Weimar and before, yet it was nonetheless a party. The reasons why this was so may help to add to our understanding of what a party is. In the same way, another question which must be asked can also extend our knowledge of a difficult matter: if the SPD in exile was not a political party in the ordinary sense of the word, what status may be ascribed to its leadership? Why was it possible for the British Government and the British Labour movement to accept that an exile like Erich Ollenhauer *was* a leading Social Democrat before 1941 and after 1945, but *not* one in the period in between?

As far as the SPD's exile status is concerned, it can be argued that by gathering and organizing a membership in England from 1939 until 1946 it was a political party, if only a very small one. As far as its leadership is concerned it can be suggested that it not only possessed the allegiance of that membership but that implicit in its treatment by various British bodies there was the acceptance that one day another German political system would exist in which a German Socialist party would play an important role. This was after all, the reason why its leadership was brought to England in the first place. What subsequent friction and hostility there was was due to other factors, but it was never suggested that a future SPD would not be a powerful force and that its leadership in exile, inasmuch as it remained unchallenged and alive, would not be of significance. Even if some believed that to have been in exile during the war would spell political death in Germany after the war, this view was based on ignorance of the leadership. Indeed those who did know what the Social Democratic leaders were demanding after 1942 were embarrassed and mistrustful of it precisely because they accepted the appeal of SPD arguments.

The evidence, then, seems to prove that what, on the face of it, may seem a story of 'frustration, quarrelling and failure' is in reality an important piece of political life which throws light on a number of interesting problems. The SPD leaders, it will be argued above, were for the most part highly successful in exile. First and foremost they enabled the SPD to survive as the political leaders of organized German labour. In so doing they had to surmount immensely difficult challenges from the British Labour movement, from the British Foreign Office, from German Communists, and even from SPD leaders in the USA who resented the forcefulness of their colleagues in London and disputed their right to the leadership of the post-war party.

Some of these challenges were met by the deployment of tactical devices which demonstrate skill of high quality. Others were overcome through luck, a factor that should not be underestimated. It was usually, however, a mixture of these two things with the addition of a very powerful factor, the historical traditions of the SPD, which brought the party in exile over the worst hurdles imaginable and turned political pariahs into forces which could not be ignored.

Bibliography

I Documentary Sources and Source Materials

(a) Unpublished archival material
The archives of the SPD, in the Friedrich-Ebert-Stiftung, Bonn, West
    Germany
Files (*Mappen*) 1-182
Nachlass Carl Severing
Nachlass Hermann Müller

The archives of the International Institute for Social History Amsterdam,
    Holland
Nachlass Grzesinski
Nachlass Otto Braun

The Public Record Office, London, England
Files of the Foreign Office, 1939–46: FO 371.C: 2298 22904 22946
23002 23008 23010 23029 23036 23094 23105 24100 24101 24289
24299 24326 24327 24362 24364 24386 24387 24388 24389 24392
24409 24412 24420 24424 24437 24418 24419 25106 25210 25247
25248 25253 26520 26531 26532 26533 26537 26546 26549 26553
26554 26558 26559 26560 26581 26582 26585 26590 26597 30864
30911 30928 30929 30930 30931 30936 30942 30950 30952 30958
32618 32623 34329 34366 34397 34399 34444 34445 34459 34467
34413 34414 34415 34416 34428 34429 34430 34436 34437 34438
34439 34454 34456 34457 34458 34459 34460 34461 34462 34463
34467 34472 34474 34475 34476 34481 39061 39062 39063 39066
39119 39120 39184 46719 46826 46827 46828 46829 46911 46969

The archives of the British Labour party, London
Middleton Papers
International Subcommittee Minutes and Documents
International Department Correspondence 1932–46

The Wiener Library, London
Paul Anderson papers

University College, Oxford
Attlee papers 1939–46 (Boxes 2, 4, 6, 7, 8, 9, 14, 15, 16, 44)

(b) Interviews
Heinrich Fraenkel, July 1973, London
Fritz Heine, August 1975, August 1976, July 1977, Bad Münstereifel
Professor Gerhard Leibholz, August 1977, Göttingen
Lord Noel-Baker, November 1977, London
Susanne Miller, July 1978, Bonn
Lord Sherfield (Roger Makins) May 1979, London
Sir Frank Roberts, June 1979

(c) Published Documents
Labour party conference reports 1938–46
Die Wiedergeburt der deutschen Sozialdemokratie. Bericht über die sozialdemokratischen Parteikonferenz von Hanover, 5. bis 7. Oktober 1945
House of Commons Debates 1938–46
House of Lords Debates 1938–46
Excerpts from:
Sozialistische Mitteilungen, SPD archives, Bonn (Microfilm)
Die Zeitung, PRO, London
The Bystander, 3 July 1940, Transport House, London
Evening News, 9 January 1941
Sunday Times, 9 and 23 November 1941
The Times
The Daily Herald
The Sunday Dispatch

II Secondary Sources

Abendroth, W., *A Short History of the European Working Class* (London, 1972).
——, 'Der Widerstand der Arbeiterbewegung' in *Deutscher Widerstand,* ed. E. Weick (Heidelberg, 1967).
——, *Aufstieg und Krise der SPD* (Frankfurt, 1964).
Acheson, D., *Sketches from life* (London, 1961).
Acland, R. and Fraenkel, H., *The Winning of the Peace* (London, 1942).
Adenauer, K., *Erinnerungen 1945-1953* (Frankfurt, 1967).
Almond, G. A. (ed.), *The Struggle for Democracy in Germany* (2nd edn., New York, 1965).
Balfour, M., *Propaganda in War 1939-1945* (London, 1979).
Binder, D., *The Other German* (Washington, 1975).
Birley, R., *The German Problem and the Responsibility of Britain* (London, 1947).
Boberach, H., *Meldungen aus dem Reich 1933-1945* (Müchen, 1968).
Boyle, A., *The Climate of Treason* (2nd edn., London, 1980).
Bracher, K. D., *The German Dictatorship* (London, 1973).
——, *Die Auflösung der Weimarer Republik* (Villingen, 1964).
Brandt, W., *In Exile* (London, 1971).
Braunthal, J., *Need Germany Survive?* (London, 1943).

Briggs, A., *The History of Broadcasting in the United Kingdom, Vol. 3, The War of the Words* (London, 1970).
Bruegel, J. W., *Czechoslovakia Before Munich* (Cambridge, 1973).
Bullock, A., *Ernest Bevin*, Vol. 2 (London, 1967).
—, *Hitler: A study in Tyranny* (London, 1971).
Burridge, T. D., *British Labour and Hitler's War* (London, 1976).
Carsten, F. L., *The Reichswehr and Politics 1918-1933* (Oxford, 1966).
Churchill, W. S., *The Second World War*, 6 vols. (London, 1948-1954).
Colvin, I., *Vansittart in Office* (London, 1965).
Cowling, M., *The Impact of Hitler* (Chicago, 1975).
Dalton, H., *Hitler's War* (London, 1940).
—, *The Fateful Years* (London, 1957).
Delmer, S., *Trail Sinister*, Vol. 1 and Vol. 2 (London, 1961, 1962).
Dilks, D. (ed.), *The Diaries of Sir A. Cadogan* (London, 1971).
Dimitroff, G., *The United Front* (London, 1938).
Domarus, M., *Hitler 1932-1935*, 2 vols. (München, 1965).
Dowe, D. and Klotzbach, K., *Programmatische Dokumente der deutschen Sozialdemokratie* (Berlin, 1973).
Drechsler, H., *Die Sozialisitische Arbeiterpartei Deutschlands* (Meisenheim, 1965).
Duhnke, H., *Die KPD von 1933 bis 1945* (Köln, 1972).
Edinger, L., *German Exile Politics* (California, 1956).
—, *Kurt Schumacher* (Stanford, 1965).
Foot, M. R. D., *The Resistance* (London, 1976).
Fraenkel, H., *The German People versus Hitler* (London, 1940).
—, *Help Us Germans to Beat Hitler* (London, 1941).
—, *Germany's Road to Democracy* (London, 1943).
—, *Lebewohl Deutschland* (Hanover, 1960).
—, *The Other Germany* (London, -).
Geyer, C., *Die Partei der Freiheit* (Paris, 1939).
Glees, A., 'Albert C. Grzesinski and the Politics of Prussia', *English Historical Review*, LXXXIX (October, 1974), 814-34.
Gollancz, V., *Shall our Children live or die?* (London, 1942).
Gottfurcht, H., *Die Internationale Gewerkschaftsbewegung im Weltgeschehen* (Köln, 1963).
Grebing, H., *Die Geschichte der deutschen Arbeiterbewegung* (München, 1966).
Gross, B., 'The German Communists' united front and popular front ventures' in *The Comintern*, ed. M. Drachovitch (New York, 1966).
Grosser, A., *Germany in our Time* (London, 1974).
Grossman, K., *Emigration* (Frankfurt, 1969).
Grzesinski, A. C., *Inside Germany* (New York, 1939).
Hagen, P., *Will Germany Crack?* (London, 1943).
Harprecht, K., *Willy Brandt* (München, 1970).
Harris, J., *William Beveridge, a biography* (Oxford), 1977).
Hearndon, A., *The British in Germany* (London, 1978).
Heine, F., *Presse, Politik und Werbung* (Berlin, 1964).

Hermsdorf, H., *Erich Ollenhauer, Mensch und Leistung, Ansprache zum 10. Todestag* (Bonn, 1974).

Hiller, K., *The Problem of Constitution (sic)* (London, 1945).

— (ed.), *After Nazism—Democracy?* (London, 1945).

Hoegner, W., *Flucht vor Hitler* (2nd ed., Munich, 1978).

Jasper, R. C. D., *G. Bell Bishop of Chichester* (London, 1967).

Jones, R. V., *Most Secret War* (London, 1978).

Kaden, A., *Einheit oder Freiheit* (Hanover, 1964).

Kirkpatrick, I., *The Inner Circle* (London, 1959).

Kettenacker, L. (ed.), *Das Andere Deutschland* (Stuttgart, 1977).

Klotzbach, K., *Bibliographie zur Geschichte der deutschen Arbeiterbewegung 1914-1945* (Bonn, 1974).

Kohn, H., *Burger Vieler Welten* (New York, 1964).

Krisch, H., *German Politics under Soviet Administration* (Columbia, 1974).

Lange, D., 'Die Haltung des SPD Parteivorstandes bei Ausbruch des zweiten Welkreiges', *Zeitschrift für Geschichtswissenschaft*, (DDR) XII (1964), 948-67.

—, 'Der Faschistische Überfall auf die Sowjetunion und die Haltung emigrierter deutscher Socialdemokratisches Führer', *Zeitschrift für Geschichtswissenschaft*, (DDR) XIV (1966), 542-67.

Laski, H. J., *The Germans—Are they Human?* (London, 1941).

Link, W., *Die Geschichte der internationalen Jugendbendes und des internationalen Sozialistischen Kampfbundes* (Meisenheim, 1964).

—, 'German political refugees—the US during World War 2' in *German Democracy and the Triumph of Hitler*, eds. E. Matthias and A. J. Nicholls (London, 1971).

Löwenthal, R. (ed.), *Die Zweite Republik* (Stuttgart, 1974).

Ludlow, P., 'The unwinding of appeasement' in *Das Andere Deutschland*, ed. L. Kettenacker (Stuttgart, 1977).

Mason, T., *Arbeiterklasse und Volksgemeinschaft* (Düsseldorf, 1975).

Matthias, E., *Sozialdemokratie und Nation* (Stuttgart, 1952).

— (ed.), *Mit dem Gesicht nach Deutschland* (Düsseldorf, 1968).

—, and Nicholls, A. J. (eds.), *German Democracy and the Triumph of Hitler* (London, 1971).

Menne, B., *The Strange Case of Dr. Brüning* (London, 1942).

Middlemas, K., *The Diplomacy of Illusion* (London, 1972).

Miller, W., *Burgfrieden und Klassenkampf* (Düsseldorf, 1974).

Mommsen, H. (ed.), *Sozialdemokratie zwischen Klassenbewegung und Volkspartei* (Frankfurt, 1974).

Moore, B. Jnr, *Injustice* (London, 1978).

Müssener, H., *Exil in Schweden* (Müchen, 1974).

Nicholls, J., *Weimar and the Rise of Hitler* (London, 1970).

— and Wheeler-Bennett, Sir J., *The Semblance of Peace* (London, 1972).

Niethammer, L., Borsdorf, U., and Brandt P. (eds.), *Arbeiterinitiative 1945* (Wuppertal, 1976).

Paetel, K. O., *Versuchung oder Chance?* (Göttingen, 1965).

Page, B., *et al.*, *Philby* (London, 1977).

Pelling, H., *Britain and the Second World War* (London, 1972).
Persico, J., *Piercing the Reich* (New York, 1979).
Philby, H. A. R., *My Silent War* (London, 1968).
Picker, H., *Hitler Tischgespräche* (München, 1968).
Pirker, T., *Die SPD nach Hitler* (München, 1965).
Prittie, T., *Germany Divided* (London, 1961).
—, *Adenauer, a study in fortitude* (London, 1971).
Radkau, J., *Die deutsche Emigration in den USA* (Düsseldorf, 1971).
Rauschning, H., *Hitler Speaks* (London, 1939).
Reed, B. and Williams, G., *Denis Healey and the Politics of Power* (London, 1971).
Röder, W., *Die deutschen sozialistischen Exilgruppen in Grossbritannien* (Hanover, 1969a).
—, 'Deutschlandpläne der Sozialdemokratischen Emigration in Grossbritannien 1942–45', *Vierteljahrescheft für Zeitgeschichte*, 17 (1969b), 72–86.
Rohe, K., *Das Reichsbanner Schwarz-Rot-Gold* (Düsseldorf, 1966).
Rose, N., *Vansittart, Study of a Diplomat* (London, 1978).
Rothfels, H., tr. Wilson, L., *The German Opposition to Hitler* (London, 1961).
Rudzio, W., 'Export englischer Demokratie?', *Vierteljahrescheft für Zeitgeschichte*, 17(1969), Dokumentation.
Ryder, A. J., *The German Revolution of 1918* (Cambridge, 1967).
Sauer, P., *Württemberg in der Zeit des National-Sozialismus* (Ulm, 1975).
Schellinger, H. K., *The SPD in the Bonn Republic* (Den Haag, 1968).
Scheurig, B., *Freies Deutschland* (München, 1960).
Schlotterbeck, F., *The Darker the Night, the Brighter the Stars* (London, 1947).
Schorske, C. E., *German Social Democracy 1905–1917* (New York, 1972).
Schulz, G., *Revolutions and Peace Treaties 1917–1920* (London, 1972).
Seale, P. and McConville, M., *Philby: the long road to Moscow* (London, 1978).
Seydlitz, W. von, *Stalingrad, Konflikt und Konsequenz* (2nd edn., Hamburg, 1977).
Smith, R. H., *The OSS* (New York, 1973).
Stampfer, F., *Erfahrungen und Erkenntnisse* (Köln, 1957).
Strang, Lord, *The Foreign Office* (London, 1955).
Strasser, O., *Germany Tomorrow* (London, 1940).
—, *Exil* (München, 1958).
Stuart, Sir C., *The Secrets of Crewe House* (London, 1920).
—, *Opportunity Knocks Once* (London, 1952).
Taylor, A. J. P., *The Origins of the Second World War* (2nd edn., London, 1963).
Thyssen, F., *I paid Hitler* (London, 1939).
Trevor-Roper, H. R. (ed.), *Hitler's Tabletalk* (London, 1953).
—, *The Philby Affair* (London, 1968).
Tutas, H. E., *Nationalsozialismus und Exil* (München, 1975).

Vansittart, R., *Black Record* (London, 1941).
—, Lord, *The Mist Procession* (London, 1958).
Walter, H. A., *Deutsche Exilliteratur*, vols. 1, 2, 7 (Darmstadt, 1974).
Wasserstein, B., *Britain and the Jews of Europe 1939–1945* (Oxford, 1979).
Wehler, H. U., *Das deutsche Kaiserreich 1871–1918* (Göttingen, 1973).
Weinert, E., *Das Nationalkomitee 'Freies Deutschland'* (East Berlin, 1957).
Wheeler, R. F., *USPD und Internationale* (Frankfurt, 1975).
Wheeler-Bennett, Sir J., *Nemesis of Power* (London, 1964).
—, *Special Relationships* (London, 1975).
Woodward, Sir L., *British Foreign Policy in the Second World War* (London, 1962).
—, *British Foreign Policy in the Second World War*, Vol. 5 (London, 1976).
Zeman, Z. A. B., *Nazi Propaganda* (Oxford, 1964).

Index